# SEMI-SIMPLE LIE ALGEBRAS
# AND
# THEIR REPRESENTATIONS

# SEMI-SIMPLE LIE ALGEBRAS
# AND
# THEIR REPRESENTATIONS

## ROBERT N. CAHN

*Lawrence Berkeley National Laboratory*
*Berkeley, California*

**DOVER PUBLICATIONS, INC.**
Mineola, New York

*Bibliographical Note*

This Dover edition, first published in 2006, is a slightly corrected and unabridged republication of the work originally published by Benjamin Cummings Publishing Company, Inc., Menlo Park, CA, in 1984. A new Preface has been specially prepared for this edition.

*International Standard Book Number: 0-486-44999-8*

Manufactured in the United States of America
Dover Publications, Inc., 31 East 2nd Street, Mineola, N.Y. 11501

# Preface to the Dover Edition

By coincidence, this book first appeared just as the explosion of string theory began, in 1984, with the demonstration by Michael Green and John Schwarz that $SO(32)$ or $E_8 \times E_8$ might fulfill Einstein's dream of unifying the forces of Nature. The fervor this engendered has not abated, despite the passage of time. The more modest hope that the non-gravitational forces might be unified through $SU(5)$, $SO(10)$, $E_6$, or the like, persists as well. By coincidence, too, the two decades since the book was published correspond neatly to two decades that began with the pursuit of the Superconducting Supercollider and are ending with the completion of its surrogate, the Large Hadron Collider. The data obtained at the LHC will undoubtedly tell us something, one way or another, about our fond hopes for unification.

The reprinting of this book has provided the opportunity to correct several typographical errors, but aside from these, the text remains the same. There is no pretense of rigor or originality. The goal is simply to provide a relatively painless introduction to the subject suitable for the mathematically inclined, but without requiring any mathematical sophistication. Despite the abstract nature of the topic, the intent is to provide a Practical Guide. Whether this will be a Practical Guide to the deepest mysteries of the Universe time will tell.

R.N.C.

Berkeley
September 2005

# Preface

Particle physics has been revolutionized by the development of a new "paradigm", that of gauge theories. The SU(2) x U(1) theory of electroweak interactions and the color SU(3) theory of strong interactions provide the present explanation of three of the four previously distinct forces. For nearly ten years physicists have sought to unify the SU(3) x SU(2) x U(1) theory into a single group. This has led to studies of the representations of SU(5), O(10), and $E_6$. Efforts to understand the replication of fermions in generations have prompted discussions of even larger groups.

The present volume is intended to meet the need of particle physicists for a book which is accessible to non-mathematicians. The focus is on the semi-simple Lie algebras, and especially on their representations since it is they, and not just the algebras themselves, which are of greatest interest to the physicist. If the gauge theory paradigm is eventually successful in describing the fundamental particles, then some representation will encompass all those particles.

The sources of this book are the classical exposition of Jacobson in his *Lie Algebras* and three great papers of E.B. Dynkin. A listing of the references is given in the Bibliography. In addition, at the end of each chapter, references are given, with the authors' names in capital letters corresponding to the listing in the bibliography.

The reader is expected to be familiar with the rotation group as it arises in quantum mechanics. A review of this material begins the book. A familiarity with SU(3) is extremely useful and this is reviewed as well. The structure of semi-simple Lie algebras is developed, mostly heuristically, in Chapters III - VII, culminating with the introduction of Dynkin diagrams. The classical Lie algebras are presented in Chapter VIII and the exceptional ones in Chapter IX. Properties of representations are explored in the next two chapters. The Weyl group is developed in Chapter XIII and exploited in Chapter XIV in the proof of Weyl's dimension formula. The final three chapters present techniques for three practical tasks: finding the decomposition of product representations, determining the subalgebras of a simple algebra, and establishing branching rules for representations. Although this is a book intended for physicists, it contains almost none of the particle physics to which it is germane. An elementary account of some of this physics is given in H. Georgi's title in this same series.

This book was developed in seminars at the University of Michigan and the University of California, Berkeley. I benefited from the students in those seminars, especially H. Haber and D. Peterson in Ann Arbor and S. Sharpe in Berkeley. Sharpe, and H.F. Smith, also at Berkeley, are responsible for many improvements in the text. Their assistance is gratefully acknowledged.

# Table of Contents

# I. SU(2)

A geometrical vector in three-dimensional space can be represented by a column vector whose entries are the x, y, and z components of the vector. A rotation of the vector can be represented by a three-by-three matrix. In particular, a rotation by $\phi$ about the z-axis is given by

$$\begin{bmatrix} \cos\phi & -\sin\phi & 0 \\ \sin\phi & \cos\phi & 0 \\ 0 & 0 & 1 \end{bmatrix}. \tag{I.1}$$

For small rotations,

$$\begin{bmatrix} \cos\phi & -\sin\phi & 0 \\ \sin\phi & \cos\phi & 0 \\ 0 & 0 & 1 \end{bmatrix} \approx I - i\phi T_z, \tag{I.2}$$

where $T_z$ is the matrix

$$\begin{bmatrix} 0 & -i & 0 \\ i & 0 & 0 \\ 0 & 0 & 0 \end{bmatrix}. \tag{I.3}$$

In a similar fashion we find $T_x$ and $T_y$:

$$T_x = \begin{bmatrix} 0 & 0 & 0 \\ 0 & 0 & -i \\ 0 & i & 0 \end{bmatrix}, \qquad T_y = \begin{bmatrix} 0 & 0 & i \\ 0 & 0 & 0 \\ -i & 0 & 0 \end{bmatrix}. \tag{I.4}$$

By direct computation we find that the finite rotations are given as exponentials of the matrices $T_x$, $T_y$, and $T_z$. Thus we have

$$\exp(-i\phi T_z) = \begin{bmatrix} \cos\phi & -\sin\phi & 0 \\ \sin\phi & \cos\phi & 0 \\ 0 & 0 & 1 \end{bmatrix}. \tag{I.5}$$

The product of two rotations like $\exp(-i\theta T_y)\exp(-i\phi T_z)$ can always be written as a single exponential, say $\exp(-i\alpha \cdot T)$ where $\alpha \cdot T = \alpha_x T_x + \alpha_y T_y + \alpha_z T_z$. Suppose we set $\exp(-i\alpha \cdot T)\exp(-i\beta \cdot T) = \exp(-i\gamma \cdot T)$ and try to calculate $\gamma$ in terms of $\alpha$ and $\beta$. If we expand the exponentials we find

$$[1 - i\alpha \cdot t - \tfrac{1}{2}(\alpha \cdot t)^2 + \cdots][1 - i\beta \cdot t - \tfrac{1}{2}(\beta \cdot t)^2 + \cdots]$$

$$= \left[1 - i(\alpha + \beta) \cdot t - \tfrac{1}{2}((\alpha + \beta) \cdot t)^2 - \tfrac{1}{2}[\alpha \cdot t, \beta \cdot t] + \cdots\right]$$

$$= \exp\{-i(\alpha + \beta) \cdot t - \tfrac{1}{2}[\alpha \cdot t, \beta \cdot t] + \cdots\}. \tag{I.6}$$

To this order in the expansion, to calculate $\gamma$ we need to know the value of the commutators like $[T_x, T_y]$, but not ordinary products like $T_x T_y$. In fact, this is true to all orders (and is known as the Campbell-Baker-Hausdorff theorem[1]). It is for this reason that we can learn most of what we need to know about Lie groups by studying the commutation relations of the generators (here, the $T$'s). By direct computation we can find the commutation relations for the $T$'s:

$$[T_x, T_y] = iT_z, \quad [T_y, T_z] = iT_x, \quad [T_z, T_x] = iT_y. \tag{I.7}$$

These commutation relations which we obtained by considering geometrical rotations can now be used to form an abstract **Lie algebra**. We suppose there are three quantities $t_x$, $t_y$, and $t_z$ with a **Lie product** indicated by $[ \, , \, ]$

$$[t_x, t_y] = it_z, \quad [t_y, t_z] = it_x, \quad [t_z, t_x] = it_y. \tag{I.8}$$

We consider all linear combinations of the $t$'s and make the Lie product linear in each of its factors and anti-symmetric:

$$[a \cdot t + b \cdot t, c \cdot t] = [a \cdot t, c \cdot t] + [b \cdot t, c \cdot t], \tag{I.9}$$

$$[a \cdot t, b \cdot t] = - [b \cdot t, a \cdot t]. \tag{I.10}$$

It is easy to show that the **Jacobi identity** follows from Eq. (I.8):

$$[a \cdot t, [b \cdot t, c \cdot t]] + [b \cdot t, [c \cdot t, a \cdot t]] + [c \cdot t, [a \cdot t, b \cdot t]] = 0. \tag{I.11}$$

When we speak of the abstract Lie algebra, the product $[a \cdot t, b \cdot t]$ is not to be thought of as $a \cdot t \, b \cdot t - b \cdot t \, a \cdot t$, since the product $a \cdot t b \cdot t$ has not been defined. When we represent the algebra by matrices (as we did at the outset), then of course the ordinary product has a well-defined meaning. Nevertheless, by custom we often refer to the Lie product as a commutator.

The abstract Lie algebra derived above from the rotation group displays the features which define Lie algebras in general. A Lie algebra is a vector space, $L$, (above, the linear combinations of the $t$'s) together with a bilinear operation (from $L \times L$ into $L$ ) satisfying

$$[x_1 + x_2, y] = [x_1, y] + [x_2, y] \ , \qquad\qquad x_1, x_2, y \in L$$

$$[ax, y] = a\,[x, y] \ , \qquad\qquad a \in F, \ x, y \in L$$

$$[x, y] = -\,[y, x] \ , \qquad\qquad x, y \in L$$

$$0 = [x, [y, z]] + [y, [z, x]] + [z, [x, y]] \ , \qquad x, y, z \in L \ . \qquad (I.12)$$

Here $F$ is the field over which $L$ is a vector space. We shall always take $F$ to be the field of real numbers, $\mathcal{R}$, or the field of complex numbers, $\mathcal{C}$.

Having motivated the formal definition of a Lie algebra, let us return to the specific example provided by the rotation group. We seek the representations of the Lie algebra defined by Eq. (I.8). By a **representation** we mean a set of linear transformations (that is, matrices) $T_x$ , $T_y$ , and $T_z$ with the same commutation relations as the $t$'s. The $T$'s of Eqs. (I.3) and (I.4) are an example in which the matrices are $3 \times 3$ and the representation is said to be of dimension three.

We recall here the construction which is familiar from standard quantum mechanics texts. It is convenient to define

$$t_+ = t_x + it_y \ , \qquad t_- = t_x - it_y \ , \qquad\qquad (I.13)$$

so that the commutation relations become

$$[t_z, t_+] = t_+ \ , \quad [t_z, t_-] = -t_- \ , \quad [t_+, t_-] = 2t_z \ . \qquad (I.14)$$

We now suppose that the $t$'s are to be represented by some linear transformations: $t_x \rightarrow T_x, t_y \rightarrow T_y, t_z \rightarrow T_z$. The $T$'s act on some vector space, $V$. We shall in fact construct this space and the $T$'s directly. We start with a single vector, $v_j$ and define the actions of $T_z$ and $T_+$ on it by

$$T_z v_j = j v_j \ , \qquad T_+ v_j = 0 \ . \qquad\qquad (I.15)$$

Now consider the vector $T_- v_j$. This vector is an eigenvector of $T_z$ with eigenvalue $j - 1$ as we see from

$$T_z T_- v_j = (T_- T_z - T_-)v_j = (j-1)T_- v_j \ . \tag{I.16}$$

Let us call this vector $v_{j-1} \equiv T_- v_j$. We proceed to define additional vectors sequentially:

$$v_{k-1} = T_- v_k \ . \tag{I.17}$$

If our space, $V$, which is to consist of all linear combinations of the $v$'s, is to be finite dimensional this procedure must terminate somewhere, say when

$$T_- v_q = 0 \ . \tag{I.18}$$

In order to determine $q$, we must consider the action of $T_+$. It is easy to see that $T_+ v_k$ is an eigenvector of $T_z$ with eigenvalue $k+1$. By induction, we can show that $T_+ v_k$ is indeed proportional to $v_{k+1}$. The constant of proportionality may be computed:

$$
\begin{aligned}
r_k v_{k+1} &= T_+ v_k \\
&= T_+ T_- v_{k+1} \\
&= (T_- T_+ + 2T_z)v_{k+1} \\
&= [r_{k+1} + 2(k+1)]v_{k+1} \ .
\end{aligned}
\tag{I.19}
$$

This recursion relation for $r_k$ is easy to satisfy. Using the condition $r_j = 0$, which follows from Eq. (I.15), the solution is

$$r_k = j(j+1) - k(k+1). \tag{I.20}$$

Now we can find the value of $q$ defined by Eq. (I.18):

$$T_+T_-v_q = 0$$

$$= (T_-T_+ + 2T_z)v_q$$

$$= [j(j+1) - q(q+1) + 2q]v_q \ . \tag{I.21}$$

There are two roots, $q = j + 1$, and $q = -j$. The former is not sensible since we should have $q \leq j$. Thus $q = -j$, and $2j$ is integral.

In this way we have recovered the familiar representations of the rotation group, or more accurately, of its Lie algebra, Eq. (I.14). The eigenvalues of $T_z$ range from $j$ to $-j$. It is straightforward to verify that the **Casimir operator**

$$T^2 = T_x^2 + T_y^2 + T_z^2$$

$$= T_z^2 + \tfrac{1}{2}(T_+T_- + T_-T_+) \ , \tag{I.22}$$

has the constant value $j(j+1)$ on all the vectors in $V$:

$$T^2 v_k = [k^2 + \tfrac{1}{2}(r_{k-1} + r_k)]v_k$$

$$= j(j+1)v_k \ . \tag{I.23}$$

The $2j + 1$ dimensional representation constructed above is said to be **irreducible**. This means that there is no proper subspace of $V$ (that is, no subspace except $V$ itself and the space consisting only of the zero vector) which is mapped into itself by the various $T$'s. A simple example of a **reducible** representation is obtained by taking two irreducible representations on the space $V_1$ and $V_2$, say, and forming the space $V_1 \oplus V_2$. That is, the vectors, $v$, in $V$ are of the form $v = v_1 + v_2$, with $v_i \in V_i$. If $t_z$ is represented by $T_z^1$ on $V_1$ and by $T_z^2$ on $V_2$, we take the representation of $t_z$ on $V$ to be $T_z(v_1 + v_2) = T_z^1 v_1 + T_z^2 v_2$, and so on for the other components. The subspaces $V_1$ and $V_2$ are **invariant** (that is, mapped into themselves) so the representation is reducible.

A less trivial example of a reducible representation occurs in the "addition of angular momentum" in quantum mechanics. Here we combine two irreducible representations by forming the product space $V = V_1 \otimes V_2$. If the vectors $u_{1m}$ and $u_{2n}$ form bases for $V_1$ and $V_2$ respectively, a basis for $V$ is given by the quantities $u_{1m} \otimes u_{2n}$. We define the action of the $T$'s on $V$ by

$$T_z(u_{1m} \otimes u_{2n}) = (T_z^1 u_{1m}) \otimes u_{2n} + u_{1m} \otimes (T_z^2 u_{2n}) \,, \tag{I.24}$$

etc. If the maximum value of $T_z^1$ on $V_1$ is $j_1$ and that of $T_z^2$ on $V_2$ is $j_2$, there is an eigenvector of $T_z = T_z^1 + T_z^2$ with eigenvalue $j_1 + j_2$. By applying $T_- = T_-^1 + T_-^2$ repeatedly to this vector, we obtain an irreducible subspace, $U_{j_1+j_2}$, of $V_1 \otimes V_2$. On this space, $T^2 = (j_1+j_2)(j_1+j_2+1)$. Indeed, we can decompose $V_1 \otimes V_2$ into a series of subspaces on which $T^2$ takes the constant value $k(k+1)$ for $|j_1 - j_2| \leq k \leq j_1 + j_2$, that is $V_1 \otimes V_2 = U_{j_1+j_2} \oplus \ldots \oplus U_{|j_1-j_2|}$.

The representation of smallest dimension has $j = 1/2$. Its matrices are $2 \times 2$ and traceless. The matrices for $T_x, T_y,$ and $T_z$ are **hermitian** (a hermitian matrix $M$, satisfies $M_{ji}^* = M_{ij}$ where * indicates complex conjugation). If we consider the real linear combinations of $T_x, T_y,$ and $T_z$ we obtain matrices, $T$, which are traceless and hermitian. The matrices $\exp(iT)$ form a group of **unitary** matrices of determinant unity (a matrix is unitary if its adjoint - its complex conjugate transpose - is its inverse). This group is called $SU(2)$, $S$ for "special" ( determinant equal to unity), and $U$ for unitary. The rotations in three dimensions, $O(3)$, have the same Lie algebra as $SU(2)$ but are not identical as groups.

## Footnote

1. See, for example, JACOBSON, pp. 170–174.

# References

This material is familiar from the treatment of angular momentum in quantum mechanics and is presented in all the standard texts on that subject. An especially fine treatment is given in GOTTFRIED.

# Exercises

Define the standard Pauli matrices

$$\sigma_x = \begin{bmatrix} 0 & 1 \\ 1 & 0 \end{bmatrix}, \quad \sigma_y = \begin{bmatrix} 0 & -i \\ i & 0 \end{bmatrix}, \quad \sigma_z = \begin{bmatrix} 1 & 0 \\ 0 & -1 \end{bmatrix}.$$

1. Prove that $t_x \rightarrow \frac{1}{2}\sigma_x$, $t_y \rightarrow \frac{1}{2}\sigma_y$, etc. is a representation of SU(2).

2. Prove, if $\alpha \cdot \sigma = \alpha_x\sigma_x + \alpha_y\sigma_y + \alpha_z\sigma_z$, etc. then $\alpha \cdot \sigma \beta \cdot \sigma = \alpha \cdot \beta + i\alpha \times \beta \cdot \sigma$.

3. Prove that $\exp(-i\theta\sigma \cdot n/2) = \cos(\theta/2) - in \cdot \sigma \sin(\theta/2)$, where $n \cdot n = 1$.

4. Prove $\exp(-i\theta\sigma \cdot n/2)\sigma \cdot n' \exp(i\theta\sigma \cdot n/2) = \sigma \cdot n''$, where $n \cdot n = n' \cdot n' = 1$ and where $n'' = \cos\theta \; n' + n \cdot n'(1 - \cos\theta)n + \sin\theta \; n \times n'$. Interpret geometrically.

5. Prove $\exp(-i2\pi n \cdot T) = (-1)^{2j}$ where $n \cdot n = 1$ and $T^2 = j(j + 1)$.

# II. SU(3)

The preceding review of SU(2) will be central to the understanding of Lie algebras in general. As an illustrative example, however, SU(2) is not really adequate. The Lie algebra of SU(3) is familiar to particle physicists and exhibits most of the features of the larger Lie algebras that we will encounter later.

The group SU(3) consists of the unitary, three-by-three matrices with determinant equal to unity. The elements of the group are obtained by exponentiating $iM$, where $M$ is a traceless, three-by-three, hermitian matrix. There are eight linearly independent matrices with these properties.

One choice for these is the $\lambda$ matrices of Gell-Mann:

$$\lambda_1 = \begin{bmatrix} 0 & 1 & 0 \\ 1 & 0 & 0 \\ 0 & 0 & 0 \end{bmatrix}, \qquad \lambda_2 = \begin{bmatrix} 0 & -i & 0 \\ i & 0 & 0 \\ 0 & 0 & 0 \end{bmatrix}, \qquad \lambda_3 = \begin{bmatrix} 1 & 0 & 0 \\ 0 & -1 & 0 \\ 0 & 0 & 0 \end{bmatrix},$$

$$\lambda_4 = \begin{bmatrix} 0 & 0 & 1 \\ 0 & 0 & 0 \\ 1 & 0 & 0 \end{bmatrix}, \qquad \lambda_5 = \begin{bmatrix} 0 & 0 & -i \\ 0 & 0 & 0 \\ i & 0 & 0 \end{bmatrix}, \qquad \lambda_6 = \begin{bmatrix} 0 & 0 & 0 \\ 0 & 0 & 1 \\ 0 & 1 & 0 \end{bmatrix}, \quad \text{(II.1)}$$

$$\lambda_7 = \begin{bmatrix} 0 & 0 & 0 \\ 0 & 0 & -i \\ 0 & i & 0 \end{bmatrix}, \qquad \lambda_8 = \frac{1}{\sqrt{3}} \begin{bmatrix} 1 & 0 & 0 \\ 0 & 1 & 0 \\ 0 & 0 & -2 \end{bmatrix}.$$

The first three are just the Pauli matrices with an extra row and column added. The next four also have a similarity to $\sigma_x$ and $\sigma_y$. To exploit this similarity we define

$$T_x = \tfrac{1}{2}\lambda_1, \qquad T_y = \tfrac{1}{2}\lambda_2, \qquad T_z = \tfrac{1}{2}\lambda_3,$$

$$V_x = \tfrac{1}{2}\lambda_4, \qquad V_y = \tfrac{1}{2}\lambda_5,$$

$$U_x = \tfrac{1}{2}\lambda_6, \qquad U_y = \tfrac{1}{2}\lambda_7, \qquad Y = \frac{1}{\sqrt{3}}\lambda_8. \qquad \text{(II.2)}$$

There is no $U_z$ or $V_z$ because there are only two linearly independent diagonal generators. By historical tradition, they are chosen to be $T_z$ and $Y$. Just as with SU(2), it is convenient to work with the complex combinations

$$T_\pm = T_x \pm iT_y, \quad V_\pm = V_x \pm iV_y, \quad U_\pm = U_x \pm iU_y. \qquad \text{(II.3)}$$

It is straightforward to compute all the commutation relations between the eight generators. See Table II.1. We can consider these commutation relations to define the abstract Lie algebra of SU(3). That is, a representation of SU(3) is a correspondence $t_z \rightarrow T_z$, $t_+ \rightarrow T_+$, $t_- \rightarrow T_-$, $u_+ \rightarrow U_+$, etc. which preserves the commutation relations given in Table II.1. The three-by-three matrices given above form one representation, but as is well-known, there are six dimensional, eight dimensional, ten dimensional representations, etc.

# Table II.1

The SU(3) commutation relations. The label on the row gives the first entry in the commutator and the column label gives the second.

|  | $t_+$ | $t_-$ | $t_z$ | $u_+$ | $u_-$ | $v_+$ | $v_-$ | $y$ |
|---|---|---|---|---|---|---|---|---|
| $t_+$ | 0 | $2t_z$ | $-t_+$ | $v_+$ | 0 | 0 | $-u_-$ | 0 |
| $t_-$ | $-2t_z$ | 0 | $t_-$ | 0 | $-v_-$ | $u_+$ | 0 | 0 |
| $t_z$ | $t_+$ | $-t_-$ | 0 | $-\frac{1}{2}u_+$ | $\frac{1}{2}u_-$ | $\frac{1}{2}v_+$ | $-\frac{1}{2}v_-$ | 0 |
| $u_+$ | $-v_+$ | 0 | $\frac{1}{2}u_+$ | 0 | $\frac{3}{2}y - t_z$ | 0 | $t_-$ | $-u_+$ |
| $u_-$ | 0 | $v_-$ | $-\frac{1}{2}u_-$ | $-\frac{3}{2}y + t_z$ | 0 | $-t_+$ | 0 | $u_-$ |
| $v_+$ | 0 | $-u_+$ | $-\frac{1}{2}v_+$ | 0 | $t_+$ | 0 | $\frac{3}{2}y + t_z$ | $-v_+$ |
| $v_-$ | $u_-$ | 0 | $\frac{1}{2}v_-$ | $-t_-$ | 0 | $-\frac{3}{2}y - t_z$ | 0 | $v_-$ |
| $y$ | 0 | 0 | 0 | $u_+$ | $-u_-$ | $v_+$ | $-v_-$ | 0 |

The eight dimensional representation is especially interesting. It can be obtained in the following fashion. We seek a mapping of the generators $t_+$, $t_-$, $t_z$, $u_+$, *etc.* into linear operators which act on some vector space. Let that vector space be the Lie algebra, $L$, itself, that is, the space of all linear combinations of $t's$, $u's$, *etc.* With $t_z$ we associate the following linear transformation. Let $x \in L$ and take

$$x \rightarrow [t_z, x] \ . \tag{II.4}$$

We call this linear transformation $\operatorname{ad} t_z$. More generally, if $x$, $y \in L, \operatorname{ad} y(x) = [y, x]$ .

Now the mapping $y \rightarrow \operatorname{ad} y$ is just what we want. It associates with each element $y$ in the Lie algebra, a linear tranformation, namely $\operatorname{ad} y$. To see that this is a representation, we must show it preserves the commutation relations, that is, if $[x, y] = z$ it must follow that $[\operatorname{ad} x, \operatorname{ad} y] = \operatorname{ad} z$. (It is worth noting here that the brackets in the first relation stand for some abstract operation, while those in the second indicate ordinary commutation.) This is easy to show:

$$[\operatorname{ad} x, \operatorname{ad} y] \, w = [x, [y, w]] - [y, [x, w]]$$

$$= [x, [y, w]] + [y, [w, x]]$$

$$= - [w, [x, y]]$$

$$= [[x, y] \, , w] = [z, w]$$

$$= \operatorname{ad} z(w) \ . \tag{II.5}$$

In the third line the Jacobi identity was used.

This representation is eight dimensional since $L$ is eight dimensional. The operators $\operatorname{ad} x$ can be written as eight-by-eight matrices if we select a particular basis. For example, if the basis for $L$ is $t_+$, $t_-$, $t_z$, $u_+$, $u_-$, $v_+$, $v_-$, and $y$ (in that order), the matrix for $\operatorname{ad} t_+$ is found to be (using Table II.1)

$$\operatorname{ad} t_+ = \begin{bmatrix} 0 & 0 & -1 & 0 & 0 & 0 & 0 & 0 \\ 0 & 0 & 0 & 0 & 0 & 0 & 0 & 0 \\ 0 & 2 & 0 & 0 & 0 & 0 & 0 & 0 \\ 0 & 0 & 0 & 0 & 0 & 0 & 0 & 0 \\ 0 & 0 & 0 & 0 & 0 & 0 & -1 & 0 \\ 0 & 0 & 0 & 1 & 0 & 0 & 0 & 0 \\ 0 & 0 & 0 & 0 & 0 & 0 & 0 & 0 \\ 0 & 0 & 0 & 0 & 0 & 0 & 0 & 0 \end{bmatrix}, \tag{II.6}$$

while that for $\operatorname{ad} t_z$ is

$$\operatorname{ad} t_z = \begin{bmatrix} 1 & 0 & 0 & 0 & 0 & 0 & 0 & 0 \\ 0 & -1 & 0 & 0 & 0 & 0 & 0 & 0 \\ 0 & 0 & 0 & 0 & 0 & 0 & 0 & 0 \\ 0 & 0 & 0 & -\frac{1}{2} & 0 & 0 & 0 & 0 \\ 0 & 0 & 0 & 0 & \frac{1}{2} & 0 & 0 & 0 \\ 0 & 0 & 0 & 0 & 0 & \frac{1}{2} & 0 & 0 \\ 0 & 0 & 0 & 0 & 0 & 0 & -\frac{1}{2} & 0 \\ 0 & 0 & 0 & 0 & 0 & 0 & 0 & 0 \end{bmatrix}. \tag{II.7}$$

Both $\operatorname{ad} t_z$ and $\operatorname{ad} y$ are diagonal. Thus if $x = a t_z + b y$, then $\operatorname{ad} x$ is diagonal. Explicitly, we find

$$\operatorname{ad} x = \begin{bmatrix} a & 0 & 0 & 0 & 0 & 0 & 0 & 0 \\ 0 & -a & 0 & 0 & 0 & 0 & 0 & 0 \\ 0 & 0 & 0 & 0 & 0 & 0 & 0 & 0 \\ 0 & 0 & 0 & -\frac{1}{2}a + b & 0 & 0 & 0 & 0 \\ 0 & 0 & 0 & 0 & \frac{1}{2}a - b & 0 & 0 & 0 \\ 0 & 0 & 0 & 0 & 0 & \frac{1}{2}a + b & 0 & 0 \\ 0 & 0 & 0 & 0 & 0 & 0 & -\frac{1}{2}a - b & 0 \\ 0 & 0 & 0 & 0 & 0 & 0 & 0 & 0 \end{bmatrix}. \tag{II.8}$$

In other words, $t_+$, $t_-$, $t_z$, $u_+$, $u_-$, $v_+$, $v_-$, and $y$ are all eigenvectors of $\operatorname{ad} x$ with eigenvalues $a, -a, 0, -\frac{1}{2}a + b, \frac{1}{2}a - b, \frac{1}{2}a + b, -\frac{1}{2}a - b$, and $0$ respectively.

The procedure we have followed is central to the analysis of larger Lie algebras. We have found a two dimensional subalgebra (all linear combinations of

$t_z$ and $y$) which is **abelian** (that is, if $x$ and $y$ are in the subalgebra, $[x, y] = 0$). We have chosen a basis for the rest of the Lie algebra so that each element of the basis is an eigenvector of ad $x$ if $x$ is in the subalgebra (called the **Cartan subalgebra**). It is for this reason that we chose to work with $t_+$ and $t_-$ rather than $t_x$ and $t_y$, etc.

Once we have selected the Cartan subalgebra, $H$, the determination of the eigenvectors of ad $x$ for $x \in H$ does not depend on a specific choice of basis for $H$. That is, we could choose any two linearly independent combinations of $t_z$ and $y$ as the basis for $H$. Of course, the eigenvectors are not uniquely determined, but are determined only up to a multiplicative constant: if $u_+$ is an eigenvector of ad $x$, then so is $cu_+$, where $c$ is a number. The eigenvalues, however, are completely determined, since, for example, $u_+$ and $cu_+$ have the same eigenvalue.

These eigenvalues depend, of course, on what $x$ is. Specifically, we have

$$\text{ad}\,(at_z + by)t_+ = at_+$$

$$\text{ad}\,(at_z + by)t_- = -at_-$$

$$\text{ad}\,(at_z + by)t_z = 0t_z$$

$$\text{ad}\,(at_z + by)u_+ = (-\tfrac{1}{2}a + b)u_+$$

$$\text{ad}\,(at_z + by)u_- = (\tfrac{1}{2}a - b)u_-$$

$$\text{ad}\,(at_z + by)v_+ = (\tfrac{1}{2}a + b)v_+$$

$$\text{ad}\,(at_z + by)v_- = (-\tfrac{1}{2}a - b)v_-$$

$$\text{ad}\,(at_z + by)y = 0y \;. \tag{II.9}$$

The eigenvalues depend on $x$ in a special way: they are linear functions of $x$. We may write

$$\alpha_{u_+}(at_z + by) = -\tfrac{1}{2}a + b \;, \tag{II.10}$$

etc. The functions $\alpha_{u_+}$ are linear functions which act on elements of the Cartan subalgebra, $H$, and have values in the complex numbers. The mathematical term for a linear function which takes a vector space, $V$ (here $V$ is $H$, the Cartan subalgebra) into the complex numbers is a **functional**. The linear functionals on $V$ comprise a vector space called the **dual space**, denoted $V^*$. Thus we say that the functionals $\alpha$ lie in $H^*$. These functionals, $\alpha$, are called **roots** and the corresponding generators like $u_+$ are called **root vectors**.

The concept of a dual space may seem excessively mathematical, but it is really familiar to physicists in a variety of ways. If we consider a geometrical vector space, say the three-dimensional vectors, there is a well-defined scalar (dot) product. Thus if $\mathbf{a}$ and $\mathbf{b}$ are vectors, $\mathbf{a} \cdot \mathbf{b}$ is a number. We can think of $\mathbf{a}\cdot$ as an element in the dual space. It acts on vectors to give numbers. Moreover, it is linear: $\mathbf{a}{\cdot}(\mathbf{b} + \mathbf{c}) = \mathbf{a} \cdot \mathbf{b} + \mathbf{a} \cdot \mathbf{c}$. A less trivial example is the bra-ket notation of Dirac: $|\ \rangle$ represents vectors in the space $V$, $\langle|$ represents vectors in the dual space, $V^*$.

# References

SU(3) is familiar to particle physicists and is presented in many texts. Particularly notable presentations are found in GASIOROWICZ and in CARRUTHERS.

# Exercises

1. Show that $\lambda_1, \lambda_2,$ and $\lambda_3$ close among themselves under the commutation relations, that is that they generate an SU(2) subalgebra. Show the same is true for $\lambda_2, \lambda_5,$ and $\lambda_7$.

2. Show that

$$\sum_i \lambda^i_{ab} \lambda^i_{cd} = -\frac{1}{3} \sum_i \lambda^i_{ad} \lambda^i_{cb} + \frac{16}{9} \delta_{ad} \delta_{cb} \ .$$

# III. The Killing Form

A fundamental step in the analysis of Lie algebras is to establish a geometrical picture of the algebra. We shall eventually see how this geometry can be developed in terms of the roots of the algebra. Before turning to the roots, we must first define something resembling a scalar product for elements of the Lie algebra itself. We shall state our definitions for an arbitrary Lie algebra and illustrate them with SU(3).

Let $L$ be a Lie algebra and let $a, b \in L$. The **Killing form** is defined by

$$(a, b) = \text{Tr ad } a \text{ ad } b . \tag{III.1}$$

Remember that ad $a$ is an operator which acts on elements of $L$ and maps them into new elements of $L$. Thus the indicated trace can be evaluated by first taking a basis for $L$, say $x_1, x_2, \ldots$. Then we calculate for each $x_j$, the quantity $[a, [b, x_j]]$ and express the result in terms of the $x_i$'s. The coefficient of $x_j$ is the contribution to the trace. It is easy to show that the trace is independent of the choice of basis. As an example of the Killing form, consider SU(3). Using Table II.1 we see that

$$(t_z, t_z) = 3. \tag{III.2}$$

This can be calculated simply using the matrix representation of the operator $\mathrm{ad}\, t_z$, Eq. (II.7), or more tediously

$$
\begin{aligned}
[t_z, [t_z, t_z]] &= 0 \ , & [t_z, [t_z, u_+]] &= \tfrac{1}{4} u_+ \\
[t_z, [t_z, y]] &= 0 \ , & [t_z, [t_z, u_-]] &= \tfrac{1}{4} u_- \\
[t_z, [t_z, t_+]] &= t_+ \ , & [t_z, [t_z, v_+]] &= \tfrac{1}{4} v_+ \\
[t_z, [t_z, t_-]] &= t_- \ , & [t_z, [t_z, v_-]] &= \tfrac{1}{4} v_- \ .
\end{aligned}
\tag{III.3}
$$

It is easy to see that a term like $(t_z, t_+)$ must vanish. From Table II.1 we see that $\mathrm{ad}\, t_z \, \mathrm{ad}\, t_+ \ (t_z) = -t_+$ and hence gives no contribution to $(t_z, t_+)$, *etc.* If we take the Killing form between two of our basis elements, only a few are non-zero:

$$
\begin{aligned}
(t_z, t_z) &= 3 \ , & (y, y) &= 4 \ , & (t_+, t_-) &= 6 \ , \\
(v_+, v_-) &= 6 \ , & (u_+, u_-) &= 6 \ .
\end{aligned}
\tag{III.4}
$$

The Killing form is not a scalar product. In particular it is not positive definite. For example, since we are considering complex combinations of the SU(3) generators, we can calculate $(iy, iy) = -4$.

There is a scalar product associated with the Lie algebra, but it is not defined on the Lie algebra itself, but rather on the space containing the roots. We recall that the roots live in a space called $H^*$, the dual space to the Cartan subalgebra, $H$. Often we can restrict ourselves to the space $H_0^*$, the real, linear combinations of the roots.

The Killing form enables us to make a connection between the Cartan subalgebra, $H$, and its dual $H^*$.[1] If $\rho \in H^*$, there exists a unique element $h_\rho \in H$ such that for every $k \in H$,

$$\rho(k) = (h_\rho, k) . \tag{III.5}$$

This unique connection between $H$ and $H^*$ occurs not for all Lie algebras but only for the class of semi-simple Lie algebras which are the ones we shall be mostly concerned with. For semi-simple Lie algebras the Killing form, as we shall see, is non-degenerate. This means, in particular, that if $(a, b) = 0$ for every $b \in H$, then $a = 0$. More prosaically, non-degeneracy means that if $x_1, x_2 \ldots$ is a basis for $H$, then the matrix $(x_i, x_j)$ can be inverted. Thus the values of $(a, x_j)$ completely determine $a$.

This one-to-one relationship between $H$ and $H^*$ can be illustrated with SU(3). Referring to Eqs. (II.9) and (II.10), we designate three non-zero roots by

$$\alpha_1(at_z + by) = a$$

$$\alpha_2(at_z + by) = -\tfrac{1}{2}a + b$$

$$\alpha_3(at_z + by) = \tfrac{1}{2}a + b . \tag{III.6}$$

The other non-zero roots are the negatives of these. Now we determine the elements in $H$ corresponding to $\alpha_1$, $\alpha_2$, and $\alpha_3$. Each of these $h$'s is to lie in $H$ and is thus of the form

$$h_{\alpha_i} = c_i t_z + d_i y . \tag{III.7}$$

Using the previously computed values of the Killing form, Eq. (III.4), we see that

$$(h_{\alpha_i}, t_z) = 3c_i$$

$$(h_{\alpha_i}, y) = 4d_i . \tag{III.8}$$

To determine the coefficients $c_i$ and $d_i$, we combine the definition of $h_\alpha$, Eq. (III.5), with the expressions for the roots, Eq. (III.6):

$$\alpha_1(t_z) = 1 = (h_{\alpha_1}, t_z) = \quad 3c_1 \ ,$$

$$\alpha_1(y) = 0 = (h_{\alpha_1}, y) = \quad 4d_1 \ ,$$

$$\alpha_2(t_z) = -\tfrac{1}{2} = (h_{\alpha_2}, t_z) = 3c_2 \ ,$$

$$\alpha_2(y) = 1 = (h_{\alpha_2}, y) = \quad 4d_2 \ ,$$

$$\alpha_3(t_z) = \tfrac{1}{2} = (h_{\alpha_3}, t_z) = \quad 3c_3 \ ,$$

$$\alpha_3(y) = 1 = (h_{\alpha_3}, y) = \quad 4d_3 \ . \tag{III.9}$$

Thus we find the elements of $H$ which correspond to the various roots:

$$h_{\alpha_1} = \tfrac{1}{3}t_z; \quad h_{\alpha_2} = -\tfrac{1}{6}t_z + \tfrac{1}{4}y; \quad h_{\alpha_3} = \tfrac{1}{6}t_z + \tfrac{1}{4}y \ . \tag{III.10}$$

Of course, this correspondence is linear. It would have sufficed to determine $h_{\alpha_1}$ and $h_{\alpha_2}$ and then noted that since $\alpha_3 = \alpha_1 + \alpha_2, h_{\alpha_3} = h_{\alpha_1} + h_{\alpha_2}$. Indeed, using Eq. (III.10) we can find the element of $H$ which corresponds to any element of $H^*$ since such elements can be expressed in terms of, say, $\alpha_1$ and $\alpha_2$.

We are now in a position to display the previously advertised scalar product. Let $\alpha$ and $\beta$ be real linear combinations of the roots, that is, $\alpha, \beta \in H_0^*$ and let $h_\alpha$ and $h_\beta$ be the elements in $H$ associated with them according to Eq. (III.5). Then we define a product on $H_0^*$ by

$$\langle \alpha, \beta \rangle \equiv (h_\alpha, h_\beta) \ . \tag{III.11}$$

For the particular case of SU(3), using Eq. (III.4), we have

$$\langle \alpha_1, \alpha_1 \rangle = (h_{\alpha_1}, h_{\alpha_1}) = (\tfrac{1}{3} t_z, \tfrac{1}{3} t_z) = \tfrac{1}{3}$$

$$\langle \alpha_2, \alpha_2 \rangle = (h_{\alpha_2}, h_{\alpha_2}) = \tfrac{1}{3}$$

$$\langle \alpha_3, \alpha_3 \rangle = (h_{\alpha_3}, h_{\alpha_3}) = \tfrac{1}{3}$$

$$\langle \alpha_1, \alpha_2 \rangle = (h_{\alpha_1}, h_{\alpha_2}) = -\tfrac{1}{6}$$

$$\langle \alpha_1, \alpha_3 \rangle = (h_{\alpha_1}, h_{\alpha_3}) = \tfrac{1}{6}$$

$$\langle \alpha_2, \alpha_3 \rangle = (h_{\alpha_2}, h_{\alpha_3}) = \tfrac{1}{6} \ . \tag{III.12}$$

From these specific values, we can see that for SU(3), $\langle,\rangle$ provides a scalar product on the root space, $H_0^*$. Indeed, we can interpret Eq. (III.12) geometrically by representing the roots by vectors of length $1/\sqrt{3}$. The angles between the vectors are such that $\cos\theta = \pm\tfrac{1}{2}$ as shown in Fig. III.1.

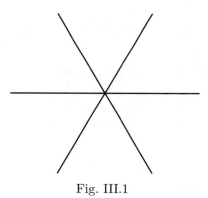

Fig. III.1

It is important to note that $\langle,\rangle$ is quite different from $(\ ,\ )$. There is no "natural" basis for the Cartan subalgebra so some of the symmetry is not apparent. Thus we found $(t_z, t_z) = 3$, but $(y, y) = 4$. Moreover, we might as well have chosen $iy$ instead of $y$ and found $(iy, iy) = -4$. There are naturally distinguished elements of $H^*$, namely the roots. As a result, the product on $H^*$ displays more clearly the symmetry of the Lie algebra.

So far we have limited our discussion to the Lie algebras of SU(2) and SU(3) (or, more precisely, their complex extensions). Let us now generalize this and explain the terms "simple" and "semi-simple".

Suppose $L$ is a Lie algebra. A **subalgebra**, M, is simply a subspace of $L$ which is closed under the Lie product. For example, $t_z, t_+$, and $t_-$ generate a subalgebra of SU(3) which is indeed isomorphic (equivalent) to SU(2). Symbolically, if $M$ is a subalgebra and $x, y \in M$, then $[x, y] \in M$. An **ideal** is a special kind of subalgebra. If $J$ is an ideal, and $x \in J$ and $y$ is any element of $L$, then $[x, y] \in J$. If $J$ were only a subalgebra instead of an ideal, we would have to restrict $y$ to be in $J$ rather than just in $L$.

As an example, consider the group U(3), the set of all three-by-three unitary matrices. We can think of its Lie algebra as being the set of all Hermitian three-by-three matrices. This is the same as for SU(3) except that the matrices need not be traceless. Thus we might take for a basis, the eight matrices displayed in Eq. (III.1), plus the three-by-three identity matrix.

Now consider the one-dimensional space spanned by the identity matrix, that is, the space given by multiples of the identity. This space, $J$, is an ideal because if $x \in J$ and $y$ is any element of the Lie algebra, $[x, y] = 0 \in J$. In fact, if we consider the space of all traceless matrices, $J'$, we see that it too is an ideal. This follows since the trace of a commutator is necessarily traceless. Thus every element in U(3) can be written as a sum of one element from $J$ and one element from $J'$. The full algebra is the sum of the two ideals.

A Lie algebra which has no ideals (except the trivial ones comprising the full algebra itself or the ideal consisting solely of 0) is called **simple**. A subalgebra in which all members commute is called **abelian**. An algebra with no abelian ideals is called **semi-simple**. Thus the Lie algebra of SU(3) is simple, while that of U(3) is neither simple nor semi-simple.

A semi-simple Lie algebra is the sum of simple ideals. Consider, for example, the five-by-five traceless hermitian matrices which are zero except for two diagonal blocks, one three-by-three and one two-by-two. Suppose we consider only matrices where each of these two blocks is separately traceless. The resulting set is a Lie algebra which can be considered the sum of two ideals, one of which is isomorphic to SU(2) and the other of which is isomorphic to SU(3). If we require only that the sum of the traces of the two diagonal blocks vanish, the resulting algebra is larger, including matrices proportional to one whose diagonal elements are $\frac{1}{3}, \frac{1}{3}, \frac{1}{3}, -\frac{1}{2}, -\frac{1}{2}$. This element and its multiples form an abelian ideal so this larger algebra $(SU(3) \times SU(2) \times U(1))$ is not semi-simple.

Because semi-simple Lie algebras are simply sums of simple ones, most of their properties can be obtained by first considering simple Lie algebras.

There is an intimate relationship between the Killing form and semi-simplicity: the Killing form is non-degenerate if and only if the Lie algebra is semi-simple. It is not hard to prove half of this fundamental theorem [2] (which is due to Cartan): if the Killing form is non-degenerate, then L is semi-simple. Suppose $L$ is not semi-simple and let $B$ be an abelian ideal. Let $b_1, b_2, \ldots$ be a basis for $B$. We can extend this to a basis for the full algebra $L$ by adding $y_1, y_2, \ldots$ where $y_i \notin B$. Now let us calculate $(b_1, a)$ where $a \in L$. First consider $[b_1, [a, b_j]]$. The inner commutator lies in $B$ since $B$ is an ideal. But then the second commutator vanishes since $B$ is abelian. Next consider $[b_1, [a, y_j]]$ The final result must lie in $B$ since $b_1 \in B$ so its expansion has no components along the $y_k$'s and along $y_j$ in particular. Thus there is no contribution to the trace. The trace vanishes and the Killing form is degenerate.

## Footnotes

1. We follow here JACOBSON, p. 110.

2. JACOBSON, pp. 69–70.

## Exercise

1. Define a bilinear form on SU(3) using the three-dimensional representation as follows. Let $x$ and $y$ be a linear combination of the matrices in Eq. (II.1) and define $((x, y)) = \operatorname{Tr} xy$. Compare this with the Killing form, i.e. $(x, y) = \operatorname{Tr} \operatorname{ad} x \operatorname{ad} y$. It suffices to consider $x$ and $y$ running over some convenient basis.

# IV. The Structure of Simple Lie Algebras

Our study of the Lie algebra of SU(3) revealed that the eight generators could be divided up in an illuminating fashion. Two generators, $t_z$ and $y$, commuted with each other. They formed a basis for the two dimensional Cartan subalgebra. The remaining generators, $u_+, u_-, v_+, v_-, t_+$, and $t_-$ were all eigenvectors of ad $t_z$ and ad $y$, that is, $[t_z, u_+]$ was proportional to $u_+$, *etc.* More generally, each of the six was an eigenvector of ad $h$ for every $h \in H$. The corresponding eigenvalue depended linearly on $h$. These linear functions on $H$ were elements of $H^*$, the dual space of $H$. The functions which gave the eigenvalues of ad $h$ were called roots and the real linear combinations of these roots formed a real vector space, $H_0^*$.

The SU(3) results generalize in the following way. Every semi-simple Lie algebra is a sum of simple ideals, each of which can be treated as a separate simple Lie algebra. The generators of the simple Lie algebra may be chosen so that one subset of them generates a commutative Cartan subalgebra, $H$. The remaining generators are eigenvectors of ad $h$ for every $h \in H$. Associated with each of these latter generators is a linear function which gives the eigenvalue. We write

$$(\mathrm{ad}\ h)e_\alpha = \alpha(h)e_\alpha \ . \tag{IV.1}$$

This is the generalization of Eq. (II.9) where we have indicated generators like $u_+, u_-$, etc., generically by $e_\alpha$.

The roots of SU(3) exemplify a number of characteristics of semi-simple Lie algebras in general. First, if $\alpha$ is a root, so is $-\alpha$. This is made explicit in Eq. (II.9), where we see that the root corresponding to $t_-$ is the negative of that corresponding to $t_+$, and so on. Second, for each root, there is only one linearly independent generator with that root. Third, if $\alpha$ is a root, $2\alpha$ is not a root.

How is the Cartan subalgebra determined in general? It turns out that the following procedure is required. An element $h \in L$ is said to be **regular** if $\mathrm{ad}\,h$ has as few zero eigenvalues as possible, that is, the multiplicity of the zero eigenvalue is minimal. In the SU(3) example, from Eq. (II.8) we see that $\mathrm{ad}\,t_z$ has a two dimensional space with eigenvalue zero, while $\mathrm{ad}\,y$ has a four dimensional space of this sort. The element $t_z$ is regular while $y$ is not. A Cartan subalgebra is obtained by finding a maximal commutative subalgebra containing a regular element. The subalgebra generated by $t_z$ and $y$ is commutative and it is maximal since there is no other element we can add to it which would not destroy the commutativity.

If we take as our basis for the algebra the **root vectors**, $e_{\alpha_1}, e_{\alpha_2} \cdots$ plus some basis for the Cartan subalgebra, say $h_1, h_2 \ldots$, then we can write a matrix representation for $\mathrm{ad}\,h$:

$$\text{ad}\,h = \begin{bmatrix} 0 & & & & & & & \\ & 0 & & & & & & \\ & & \cdot & & & & & \\ & & & 0 & & & & \\ & & & & \alpha_1(h) & & & \\ & & & & & \alpha_2(h) & & \\ & & & & & & \cdot & \\ & & & & & & & \cdot \\ & & & & & & & & \alpha_n(h) \end{bmatrix} \cdot \qquad \text{(IV. 2)}$$

From this we can see that the Killing form, when acting on the Cartan subalgebra can be computed by

$$(h_1, h_2) = \sum_{\alpha \in \Sigma} \alpha(h_1)\alpha(h_2) , \qquad \text{(IV.3)}$$

where $\Sigma$ is the set of all the roots.

We know the commutation relations between the root vectors and the members of the Cartan subalgebra, namely Eq. (IV.1). What are the commutation relations between the root vectors? We have not yet specified the normalization of the $e_\alpha$'s , so we can only answer this question up to an overall constant.

Let us use the Jacobi identity on $[e_\alpha, e_\beta]$:

$$[h, [e_\alpha, e_\beta]] = -[e_\alpha, [e_\beta, h]] - [e_\beta, [h, e_\alpha]]$$

$$= \beta(h)[e_\alpha, e_\beta] + \alpha(h)[e_\alpha, e_\beta]$$

$$= (\alpha(h) + \beta(h))[e_\alpha, e_\beta] . \qquad \text{(IV.4)}$$

This means that either $[e_\alpha, e_\beta]$ is zero, or it is a root vector with root $\alpha + \beta$, or $\alpha + \beta = 0$ , in which case $[e_\alpha, e_\beta]$ commutes with every $h$ and is thus an element of the Cartan subalgebra.

It is easy to show that $(e_\alpha, e_\beta) = 0$ unless $\alpha + \beta = 0$. This is simply a generalization of the considerations surrounding Eq. (III.3). We examine $[e_\alpha, [e_\beta, x]]$ where $x$ is some basis element of $L$, either a root vector or an element of the Cartan subalgebra. If $x \in H$, the double commutator is either zero or proportional to a root vector $e_{\alpha+\beta}$. In either case, there is no contribution to the trace. If $x$ is a root vector, say $x = e_\gamma$, the double commutator is either zero or of the form $e_{\alpha+\beta+\gamma}$, and thus does not contribute to the trace unless $\alpha + \beta = 0$.

We have seen that $[e_\alpha, e_{-\alpha}]$ must be an element of the Cartan subalgebra. We can make this more explicit with a little calculation. First we prove an important property, **invariance**, of the Killing form:

$$(a, [b, c]) = ([a, b], c) , \tag{IV.5}$$

where $a, b$, and $c$ are elements of the Lie algebra. The proof is straightforward:

$$\begin{aligned}
(a, [b, c]) &= \text{Tr ad } a \,\text{ad } [b, c] \\
&= \text{Tr ad } a \,[\text{ad } b, \text{ad } c] \\
&= \text{Tr } [\text{ad } a, \text{ad } b] \,\text{ad } c \\
&= \text{Tr ad } [a, b] \text{ad } c \\
&= ([a, b], c) .
\end{aligned} \tag{IV.6}$$

Now we use this identity to evaluate $([e_\alpha, e_{-\alpha}], h)$ where $h$ is some element of the Cartan subalgebra.

$$\begin{aligned}
([e_\alpha, e_{-\alpha}], h) &= (e_\alpha, [e_{-\alpha}, h]) \\
&= \alpha(h)(e_\alpha, e_{-\alpha}) .
\end{aligned} \tag{IV.7}$$

Both sides are linear functions of $h$. Referring to Eq. (III.5), we see that $[e_\alpha, e_{-\alpha}]$ is proportional to $h_\alpha$, where $h_\alpha$ has the property

$$(h_\alpha, k) = \alpha(k), \qquad h_\alpha, k \in H . \tag{IV.8}$$

More precisely, we have

$$[e_\alpha, e_{-\alpha}] = (e_\alpha, e_{-\alpha})h_\alpha \ . \tag{IV.9}$$

This is, of course, in accord with our results for SU(3). As an example, let $e_\alpha = u_+, e_{-\alpha} = u_-$. From Table II.1, we see that $[u_+, u_-] = 3y/2 - t_z$. From Eq. (III.4), $(u_+, u_-) = 6$, while from Eqs. (III.6), (III.10), and (II.9), we find that $h_{u_+} = y/4 - t_z/6$. Thus indeed, $[u_+, u_-] = (u_+, u_-)h_{u_+}$.

The Killing form is the only invariant bilinear form on a simple Lie algebra, up to trivial modification by multiplication by a constant. To demonstrate this, suppose that $(( \ , \ ))$ is another such form. Then

$$((h_\beta, [e_\alpha, e_{-\alpha}])) = ((h_\beta, (e_\alpha, e_{-\alpha})h_\alpha))$$

$$= (e_\alpha, e_{-\alpha})((h_\beta, h_\alpha))$$

$$= (([h_\beta, e_\alpha], e_{-\alpha}))$$

$$= (h_\beta, h_\alpha)((e_\alpha, e_{-\alpha})) \ . \tag{IV.10}$$

Thus $((h_\beta, h_\alpha))/(h_\beta, h_\alpha) = ((e_\alpha, e_{-\alpha}))/(e_\alpha, e_{-\alpha})$ and this ratio is independent of $\alpha$ as well. Thus we can write

$$\frac{((h_\beta, h_\alpha))}{(h_\beta, h_\alpha)} = k = \frac{((e_\alpha, e_{-\alpha}))}{(e_\alpha, e_{-\alpha})} \ . \tag{IV.11}$$

In a simple algebra, we can start with a single root, $\alpha$, and proceed to another root, $\beta$ such that $(h_\beta, h_\alpha) \neq 0$ and continue until we have exhausted the full set of roots, so a single value of $k$ holds for the whole algebra. Separate simple factors of a semi-simple algebra may have different values of $k$ however.

We can summarize what we have thus far learned about the structure of semi-simple Lie algebras by writing the commutation relations. We indicate the set of roots by $\Sigma$ and the Cartan subalgebra by $H$:

$$[h_1, h_2] = 0 \ , \qquad\qquad h_1, h_2 \in H$$
$$[h, e_\alpha] = \alpha(h) e_\alpha \ , \qquad\qquad \alpha \in \Sigma$$
$$[e_\alpha, e_\beta] = N_{\alpha\beta} e_{\alpha+\beta} \ , \qquad\qquad \alpha + \beta \in \Sigma$$
$$= (e_\alpha, e_{-\alpha}) h_\alpha \ , \qquad\qquad \alpha + \beta = 0$$
$$= 0 \ , \quad \alpha + \beta \neq 0 \ , \quad \alpha + \beta \notin \Sigma \ . \qquad (IV.12)$$

Here $N_{\alpha\beta}$ is some number depending on the roots $\alpha$ and $\beta$ which is not yet determined since we have not specified the normalization of $e_\alpha$.

## References

A rigorous treatment of these matters is given by JACOBSON.

## Exercise

1. Show that $at_z + by$ is almost always regular by finding the conditions on $a$ and $b$ such that it is not regular.

2. Show that invariant bilinear symmetric forms are really invariants of the Lie group associated with the Lie algebra.

# V. A Little about Representations

There is still a great deal to uncover about the structure of simple Lie algebras, but it is worthwhile to make a slight detour to discuss something about representations. This will lead to some useful relations for the adjoint representation (c.f. Eqs. (II.4) and (II.5)) and thus for the structure of the Lie algebras themselves.

The study of representations of Lie algebras is based on the simple principles discussed in Chapter I. The reason for this is that the elements $e_\alpha$, $e_{-\alpha}$, and $h_\alpha$ have commutation relations

$$[h_\alpha, e_\alpha] = \alpha(h_\alpha)e_\alpha = (h_\alpha, h_\alpha)e_\alpha = \langle \alpha, \alpha \rangle e_\alpha \ ,$$

$$[h_\alpha, e_{-\alpha}] = -\langle \alpha, \alpha \rangle e_{-\alpha} \ ,$$

$$[e_\alpha, e_{-\alpha}] = (e_\alpha, e_{-\alpha})h_\alpha \ , \tag{V.1}$$

which are just the same as those for $t_+, t_-,$ and $t_z$, except for normalization. Thus for each pair of roots, $\alpha$ , and $-\alpha$ ,there is an SU(2) we can form. What makes the

Lie algebras interesting is that the SU(2)'s are linked together by the commutation relations

$$[e_\alpha, e_\beta] = N_{\alpha\beta} e_{\alpha+\beta}, \qquad \alpha + \beta \in \Sigma . \tag{V.2}$$

We recall that a representation of the Lie algebra is obtained when for each element of the algebra we have a linear transformation (i.e. a matrix) acting on a vector space (i.e. column vectors) in a way which preserves the commutation relations. If we indicate the representation of $a$, $b$, and $c$ by $A$, $B$, and $C$, then

$$[a, b] = c \qquad \rightarrow \qquad [A, B] = C . \tag{V.3}$$

Let us continue to use this notation so that if $h_1, h_2, \ldots$ is a basis for the Cartan subalgebra $H$, we will indicate their representatives by $H_1, H_2, \ldots$ .Similarly, the representatives of $e_\alpha$ will be $E_\alpha$. The transformations $H_i$ and $E_\alpha$ act on vectors $\phi^a$ in a space, $V$. Since the $h$'s commute, so do the $H$'s. We can choose a basis for the space $V$ in which the $H$'s are diagonal (The representation is in particular a representation for the $SU(2)$ formed by $H_\alpha, E_\alpha, E_{-\alpha}$. We know how to diagonalize $H_\alpha$. But all the $H_\alpha$'s commute so we can diagonalize them simultaneously.):

$$H_i \phi^a = \lambda_i^a \phi^a . \tag{V.4}$$

The eigenvalues $\lambda_i^a$ depend linearly on the $H$'s.Thus if $h = \sum_i c_i h_i$ so that $H = \sum_i c_i H_i$, then

$$H\phi^a = \left( \sum_i c_i \lambda_i^a \right) \phi^a$$

$$\equiv M^a(h)\phi^a . \tag{V.5}$$

We can regard the eigenvalue associated with this vector, $\phi^a$, to be a linear function defined on $H$:

$$M^a \left( \sum_i c_i h_i \right) = \sum_i c_i \lambda_i^a . \tag{V.6}$$

The functions $M^a$ are called **weights**. As linear functionals on $H$, they are members of the dual space, $H^*$, just as the roots, $\alpha$, are. We shall see later that the weights can be expressed as real (in fact, rational) linear combinations of the roots. We can use the product $\langle , \rangle$ we defined on the root space also when we deal with the weights.

A simple example of Eq. (V.5) is given by the three dimensional representation of SU(3), Eqs. (II.1) and (II.2).

$$T_z = \begin{bmatrix} \frac{1}{2} & & \\ & -\frac{1}{2} & \\ & & 0 \end{bmatrix} , \qquad Y = \begin{bmatrix} \frac{1}{3} & & \\ & \frac{1}{3} & \\ & & -\frac{2}{3} \end{bmatrix} . \tag{V.7}$$

The weight vectors of the three dimensional representation are

$$\phi^a = \begin{bmatrix} 1 \\ 0 \\ 0 \end{bmatrix} , \qquad \phi^b = \begin{bmatrix} 0 \\ 1 \\ 0 \end{bmatrix} , \qquad \phi^c = \begin{bmatrix} 0 \\ 0 \\ 1 \end{bmatrix} . \tag{V.8}$$

We consider the action of $H = aT_z + bY$ on the weight vectors to find the weights:

$$
\begin{aligned}
H\phi^a &= (\tfrac{1}{2}a + \tfrac{1}{3}b)\phi^a & = M^a(at_z + by)\phi^a , \\
H\phi^b &= (-\tfrac{1}{2}a + \tfrac{1}{3}b)\phi^b & = M^b(at_z + by)\phi^b , \\
H\phi^c &= (-\tfrac{2}{3}b)\phi^c & = M^c(at_z + by)\phi^c .
\end{aligned}
\tag{V.9}
$$

The weights can be expressed in terms of the roots of SU(3), Eq. (III.6). Only two of the roots are linearly independent so we need use only two of them, say $\alpha_1$ and $\alpha_2$.

Then

$$M^a = + \tfrac{2}{3}\alpha_1 + \tfrac{1}{3}\alpha_2 \, ,$$

$$M^b = - \tfrac{1}{3}\alpha_1 + \tfrac{1}{3}\alpha_2 \, ,$$

$$M^c = - \tfrac{1}{3}\alpha_1 - \tfrac{2}{3}\alpha_2 \, . \tag{V.10}$$

In SU(2), $T_+$ and $T_-$ act as raising and lowering operators. This concept may be generalized in the following way. Suppose that $\phi^a$ is a weight vector with weight $M^a$. Then $E_\alpha \phi^a$ is a weight vector with weight $M^a + \alpha$ unless $E_\alpha \phi^a = 0$:

$$HE_\alpha \phi^a = (E_\alpha H + \alpha(h)E_\alpha)\,\phi^a$$

$$= (M^a(h) + \alpha(h))\,E_\alpha \phi^a \, . \tag{V.11}$$

Thus we can think of the $E_\alpha$ as raising operators and the $E_{-\alpha}$ as lowering operators.

If $M$ is a weight, then it lies in a string of weights $M^*$, $M^* - \alpha$, $\ldots$, $M$, $\ldots$, $M^* - q\alpha$. Let us see how $q$ is determined by $M^*$. Let $\phi_0$ be a weight vector with weight $M^*$. Then, if it is non-zero, the vector

$$(E_{-\alpha})^j \, \phi_0 = \phi_j \tag{V.12}$$

is a weight vector with weight $M^* - j\alpha$. On the other hand, $E_\alpha \phi_k$ has weight $M^* - (k-1)\alpha$, and is proportional to $\phi_{k-1}$. We can find $q$ by using the relation

$$E_{-\alpha}\phi_q = 0 \, . \tag{V.13}$$

The calculation is simplified by choosing a convenient normalization for the generators $e_\alpha$ and $e_{-\alpha} : (e_\alpha, e_{-\alpha}) = 1$. Thus

$$[e_\alpha, e_{-\alpha}] = h_\alpha \, . \tag{V.14}$$

In terms of the representation, then

$$[E_\alpha, E_{-\alpha}] = H_\alpha \ . \tag{V.15}$$

We shall also need the relation (see Eq. (III.5))

$$M(h_\alpha) = (h_M, h_\alpha) = \langle M, \alpha \rangle \ . \tag{V.16}$$

By analogy with our treatment of SU(2), we define

$$E_\alpha \phi_k = r_k \phi_{k-1} \tag{V.17}$$

and seek a recursion relation. We find

$$\begin{aligned}
E_\alpha \phi_k &= r_k \phi_{k-1} \\
&= E_\alpha E_{-\alpha} \phi_{k-1} \\
&= (E_{-\alpha} E_\alpha + H_\alpha) \phi_{k-1} \\
&= r_{k-1} \phi_{k-1} + [M^*(h_\alpha) - (k-1)\alpha(h_\alpha)] \phi_{k-1} \\
&= [r_{k-1} + \langle M^*, \alpha \rangle - (k-1)\langle \alpha, \alpha \rangle] \phi_{k-1} \ .
\end{aligned} \tag{V.18}$$

The solution to the recursion relation which satisfies $r_0 = 0$ is

$$r_k = k\langle M^*, \alpha \rangle - \tfrac{1}{2}k(k-1)\langle \alpha, \alpha \rangle \ . \tag{V.19}$$

Now from Eq. (V.13) we know that

$$E_\alpha E_{-\alpha} \phi_q = 0 = r_{q+1} \phi_q \tag{V.20}$$

so we have found $q$ in terms of $M^*$ and $\alpha$ :

$$q = \frac{2\langle M^*, \alpha \rangle}{\langle \alpha, \alpha \rangle} \ . \tag{V.21}$$

In practice, we often have a weight, $M$, which may or may not be the highest weight in the sequence $M + p\alpha, \ldots M, \ldots M - m\alpha$. We can obtain an extremely useful formula for $m - p$ by using Eq. (V.21):

$$m + p = \frac{2\langle M + p\alpha, \alpha \rangle}{\langle \alpha, \alpha \rangle},$$

$$m - p = \frac{2\langle M, \alpha \rangle}{\langle \alpha, \alpha \rangle}. \qquad (V.22)$$

As an example, let us consider the three dimensional representation of SU(3). Now suppose we wish to find the string of weights spaced by $\alpha_1$ containing the weight $M^a = \frac{2}{3}\alpha_1 + \frac{1}{3}\alpha_2$. Using the table of scalar products in Eq. (III.12), we compute

$$m - p = \frac{2\langle \frac{2}{3}\alpha_1 + \frac{1}{3}\alpha_2, \alpha_1 \rangle}{\langle \alpha_1, \alpha_1 \rangle} = 1. \qquad (V.23)$$

In fact, $m = 1$ and $p = 0$.

So important is Eq. (V.22) that it is worthwhile to pause for a geometrical interpretation. Suppose the number of weights in the string is odd. Then there is a central weight, $M_0$, such that $p = m$ and $\langle M_0, \alpha \rangle = 0$. This suggests the existence of a symmetry, a reflection which acts about the mid-point. If the full string is $M^*, M^* - \alpha, \ldots M^* - q\alpha$, this symmetry would relate the weights $M = M^* - j\alpha$ and $M' = M^* - (q - j)\alpha = M - (q - 2j)\alpha$. Using Eq. (V.22) with $p = j$ and $m = q - j$, we see that $q - 2j = 2\langle M, \alpha \rangle / \langle \alpha, \alpha \rangle$. Thus the symmetry among the weights can be expressed by

$$S_\alpha : \quad M \to M - \frac{2\langle M, \alpha \rangle}{\langle \alpha, \alpha \rangle} \alpha. \qquad (V.24)$$

It is clear that this works similarly if the string is even. In either event, $S_\alpha$ maps weights into weights.

This symmetry is called a **Weyl reflection**. If we consider elements made up of any number of reflections, these elements form a group called the **Weyl group**. The Weyl group maps weights into weights. Associated with each weight is a

**weight space** consisting of all weight vectors with a given weight. For an irreducible representation of SU(2), each weight space is one dimensional, but in general this is not so.

If we consider a string of weights, $M^*, \ldots M^* - q\alpha$, and the associated weight spaces, we can restrict our consideration to the subalgebra generated by $e_\alpha$, $e_{-\alpha}$, and $h_\alpha$. This SU(2) subalgebra is represented by $E_\alpha$, $E_{-\alpha}$, and $H_\alpha$. The representation of SU(2) on the weight spaces associated with the string is in general reducible. It contains at least one copy of the $2q + 1$ dimensional representation of SU(2). In addition, there may be other representations of lesser dimension. Each of these representations will be arranged symmetrically with respect to the reflection $S_\alpha$ with the result that $S_\alpha$ will map a weight $M$ into a weight $M'$ whose weight space has the same dimension.

The symmetry of some SU(3) representations is apparent in Fig. V.1.

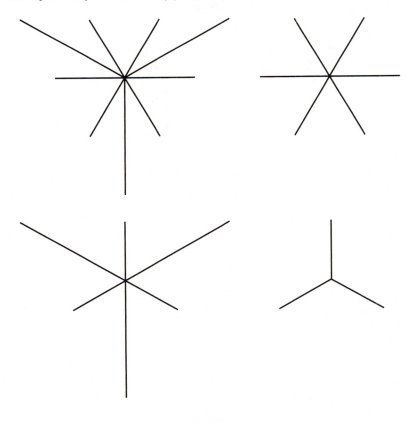

Fig. V.1

# References

The material is standard. See, for example, JACOBSON, pp. 112–119.

# Exercise

1. Find the elements of the Weyl group for SU(3) and their multiplication table.

# VI. More on the Structure of Simple Lie Algebras

In this Chapter we shall use the results on representations just obtained to learn about the algebras themselves by considering the adjoint representation. In the adjoint representation, the Lie algebra itself serves as the vector space on which the $E$'s and $H$'s act. Thus if $x$ is an element of the Lie algebra $L$, then $e_\alpha$ is represented by $E_\alpha$ where

$$E_\alpha x = \text{ad } e_\alpha(x) \ . \tag{VI.1}$$

Before studying the adjoint representation, let us first state a few properties of simple (and semi-simple) Lie algebras which may sound intuitive or obvious, but which require real mathematical proof. As is often the case, it is these innocent sounding statements which are the most difficult to prove and we omit their proofs, which may be found in standard mathematical texts.

First, if $\alpha$ is a root, then $\langle \alpha, \alpha \rangle \neq 0$. While we have asserted that $\langle \ , \ \rangle$ will become a scalar product on the root space, we have not proved it. In fact, we shall prove it later, based on the assumption that $\langle \alpha, \alpha \rangle \neq 0$.

Second, if $\alpha$ is a root, then the only multiples of $\alpha$ which are roots are $\alpha$, $-\alpha$, and 0. We can show that $-\alpha$ must be a root because $(e_\alpha, e_\beta) = 0$ unless $\alpha + \beta = 0$, and we cannot have $(e_\alpha, x) = 0$ for every $x$ in the Lie algebra, because then the Lie algebra would not be semi-simple (see Chapter III ). It might be thought that to show that $2\alpha$ is not a root would be simple, since $e_{2\alpha}$ might arise from $[e_\alpha, e_\alpha]$ which is certainly zero. However, this proves nothing since $e_{2\alpha}$ might arise from, say, $[e_{\alpha+\beta}, e_{\alpha-\beta}]$. Nevertheless, the result is true. In fact, if $\alpha, \beta$, and $\alpha + \beta$ are roots, then $[e_\alpha, e_\beta] \neq 0$.

Third, there is only one linearly independent root vector for each root. This may be stated in terms of the adjoint representation: every weight space (except the root zero space which is the Cartan subalgebra) is one dimensional.

The adjoint representation has other important properties. We know that there is no limit to the length of a string of weights for a representation: even for SU(2) we can have arbitrarily long strings, $j, j - 1 \ldots - j$. However, in the adjoint representation, a string can have no more than four weights in it. That is, a string of roots can have no more than four roots in it. We shall see that this has far-reaching consequences.

Suppose to the contrary, there is a string containing five roots which we label without loss of generality $\beta - 2\alpha, \beta - \alpha, \beta, \beta + \alpha$, and $\beta + 2\alpha$. Since $\alpha$ is a root, $2\alpha = (\beta + 2\alpha) - \beta$ is not a root, nor is $2(\beta + \alpha) = (\beta + 2\alpha) + \beta$. Thus $\beta + 2\alpha$ is in a $\beta$-string of roots with only one element. Thus from Eq. (V.22)

$$\langle \beta + 2\alpha, \beta \rangle = 0 \ . \tag{VI.2}$$

Similarly,

$$\langle \beta - 2\alpha, \beta \rangle = 0 \ . \tag{VI.3}$$

But then $\langle \beta, \beta \rangle = 0$, which is impossible. Geometrically, we see that $\beta$ is perpendicular to both $\beta + 2\alpha$ and $\beta - 2\alpha$ which is possible only if $\beta = 0$.

Now if the $\alpha$-string containing $\beta$ is four elements long, $\beta + 2\alpha, \beta + \alpha, \beta, \beta - \alpha$, then $m - p$ in Eq. (V.22) can be only $\pm 3$ or $\pm 1$. If the string is three elements long, $m - p$ can be only $\pm 2$ or $0$. If it is two elements long, $m - p$ is $\pm 1$, and if it is only one element long, $m - p$ is $0$.

We can obtain more information from the ubiquitous Eq. (V.22). Using it twice, we write

$$\frac{\langle \alpha, \beta \rangle \langle \alpha, \beta \rangle}{\langle \alpha, \alpha \rangle \langle \beta, \beta \rangle} = \tfrac{1}{4} mn \qquad (VI.4)$$

where $m$ and $n$ are integers given by the appropriate values of $m - p$. We recognize the left hand side as $\cos^2 \theta$ where $\theta$ is the angle between the vectors $\alpha$ and $\beta$. Anticipating that we really have a scalar product, we use the Schwarz inequality to assert that $mn/4$ must be less than unity unless $\alpha$ and $\beta$ are proportional. Thus $\cos^2 \theta$ can take on only the values $0$, $\tfrac{1}{4}$, $\tfrac{1}{2}$, and $\tfrac{3}{4}$.

We shall later see how this restriction of permissible angles limits the possibilities for simple Lie algebras. Indeed, we shall see that every simple Lie algebra falls either into one of four sequences of "classical" algebras or is one of the five "exceptional" Lie algebras first enumerated by Killing. Since every semi-simple Lie algebra is a sum of simple Lie algebras, this will give an exhaustive list of the semi-simple Lie algebras as well.

For the present, we pursue our analysis of the nature of roots of simple Lie algebras. First we show that every root is expressible as a linear combination of a basis set of roots with real, rational coefficients. Suppose $\alpha_1, \alpha_2 \ldots$ is a basis of roots for $H^*$. (It is not hard to show the roots span $H^*$.) Let $\beta$ be a root expressed as $\beta = \sum_i q_i \alpha_i$. Then

$$2 \frac{\langle \beta, \alpha_j \rangle}{\langle \alpha_j, \alpha_j \rangle} = \sum_i q_i \, 2 \frac{\langle \alpha_i, \alpha_j \rangle}{\langle \alpha_j, \alpha_j \rangle} \, . \qquad (VI.5)$$

This is a set of linear equations for $q_i$. All the coefficients are rational and indeed integers according to Eq. (V.22). Therefore, when we solve for the $q_i$, they will all be rational.

We can go further and show that $\langle \alpha, \beta \rangle$ is rational when $\alpha$ and $\beta$ are roots. Using Eq. (IV.3), we have

$$\langle \beta, \beta \rangle = (h_\beta, h_\beta)$$

$$= \sum_{\alpha \in \Sigma} \alpha(h_\beta)\alpha(h_\beta)$$

$$= \sum_{\alpha \in \Sigma} \langle \alpha, \beta \rangle^2 . \tag{VI.6}$$

The root $\alpha$ is in some $\beta$-string of roots, so $2\langle \alpha, \beta \rangle = (m-p)_\alpha \langle \beta, \beta \rangle$ for some integral $(m-p)_\alpha$. Thus

$$\langle \beta, \beta \rangle = \sum_{\alpha \in \Sigma} \tfrac{1}{4} \left[ (m-p)_\alpha \right]^2 \langle \beta, \beta \rangle^2,$$

$$= \left( \sum_{\alpha \in \Sigma} \tfrac{1}{4} \left[ (m-p)_\alpha \right]^2 \right)^{-1} . \tag{VI.7}$$

This shows that $\langle \beta, \beta \rangle$ is rational. Also, $\langle \alpha, \beta \rangle = (m-p)_\alpha \langle \beta, \beta \rangle / 2$ is rational. We see, then, from Eq. (VI.7) that $\langle \ , \ \rangle$ is positive definite on the space of rational linear combinations of roots. In particular, this means that $\langle \ , \ \rangle$ is a scalar product.

## References

This is standard material. See, for example, JACOBSON, pp. 112–119.

## Exercise

1. Assuming that for each root $\alpha$ there is only one linearly independent root vector, show that if $\alpha$, $\beta$, and $\alpha + \beta$ are roots, then $[e_\alpha, e_\beta] \neq 0$. *Hint:* consider the adjoint representation and then the $SU(2)$ generated by $e_\alpha, e_{-\alpha}$, and $h_\alpha$.

# VII. Simple Roots and the Cartan Matrix

The next step in analyzing the simple Lie algebras is to define an ordering among the elements in the root space, the space $H_0^*$ of real linear combinations of roots. This ordering is necessarily somewhat arbitrary: there is no natural ordering in the root space. Nevertheless, we shall see that even an arbitrarily chosen ordering can provide much useful information. Let $\alpha_1, \alpha_2 \dots \alpha_n$ be a fixed basis of roots so every element of $H_0^*$ can be written $\rho = \sum_i c_i \alpha_i$. We shall call $\rho$ **positive** $(\rho > 0)$ if $c_1 > 0$, or if $c_1 = 0$, we call $\rho$ positive if $c_2 > 0$, *etc*. If the first non-zero $c_i$ is negative we call $\rho$ negative. Clearly this ordering is possible only because we consider only real linear combinations of roots rather than the full dual space, $H^*$. We shall write $\rho > \sigma$ if $\rho - \sigma > 0$.

Given the choice of an ordered basis, we can determine which roots are positive and which are negative. A **simple root** is a positive root which cannot be written as the sum of two positive roots. Let us consider $SU(3)$ as an example.

According to Eq. (III.6), the roots are

$$\alpha_1(at_z + by) = a$$

$$\alpha_2(at_z + by) = -\tfrac{1}{2}a + b$$

$$\alpha_3(at_z + by) = \tfrac{1}{2}a + b \qquad\qquad (III.6)$$

and the negatives of these roots. Suppose we select as a basis for $H_0^*$ the roots $\alpha_1$ and $\alpha_3$, in that order. Now since $\alpha_2 = \alpha_3 - \alpha_1$, $\alpha_2$ is negative. What are the simple roots? The positive roots are $\alpha_1, -\alpha_2$, and $\alpha_3$. Now $\alpha_1 = \alpha_3 + (-\alpha_2)$ so $\alpha_1$ is the sum of two positive roots and is thus not simple. The simple roots are $-\alpha_2$ and $\alpha_3$, and $-\alpha_2 > \alpha_3$. Of course, this depends on our original ordering of the basis.

We denote the set of simple roots by $\Pi$ and the set of all roots by $\Sigma$. One very important property of the simple roots is that the difference of two simple roots is not a root at all: $\alpha, \beta \in \Pi \Rightarrow \alpha - \beta \notin \Sigma$. To see this, suppose that to the contrary $\alpha - \beta$ is a root. Then either $\alpha - \beta$ or $\beta - \alpha$ is positive. Thus either $\alpha = (\alpha - \beta) + \beta$ or $\beta = (\beta - \alpha) + \alpha$ can be written as the sum of two positive roots which is impossible for simple roots.

If $\alpha$ and $\beta$ are simple roots, then $\langle \alpha, \beta \rangle \leq 0$. This follows from Eq. (V.22) because $\beta$ is a root, but $\beta - \alpha$ is not a root. Thus in Eq. (V.22), $m = 0$, so $m - p \leq 0$.

From this result it is easy to show that the simple roots are linearly independent. If the simple roots are not linearly independent we can write an equality

$$\sum_{\alpha_i \in \Pi} a_i\alpha_i = \sum_{\alpha_j \in \Pi} b_j\alpha_j \;, \qquad\qquad (VII.1)$$

where all the $a_i$ and $b_j$ are non-negative, and no simple root appears on both sides of the equation. (If there were a relation $\sum_i c_i\alpha_i = 0$ with all positive coefficients, the roots $\alpha_i$ could not all be positive.) Now multiplying both sides of Eq. (VII.1) by $\sum_i a_i\alpha_i$,

$$\langle \sum_i a_i\alpha_i, \sum_j a_j\alpha_j \rangle = \langle \sum_i a_i\alpha_i, \sum_j b_j\alpha_j \rangle \ . \tag{VII.2}$$

The left hand side is positive since it is a square, but the right hand side is a sum of negative terms. This contradiction establishes the linear independence of the simple roots. Thus we can take as a basis for the root space the simple roots, since it is not hard to show they span the space.

We now demonstrate a most important property of the simple roots: every positive root can be written as a positive sum of simple roots. This is certainly true for the positive roots which happen to be simple. Consider the smallest positive root for which it is not true. Since this root is not simple, it can be written as the sum of two positive roots. But these are smaller than their sum and so each can, by hypothesis, be written as a positive sum of simple roots. Hence, so can their sum.

From the simple roots, we form the **Cartan matrix**, which summarizes all the properties of the simple Lie algebra to which it corresponds. As we have seen, the dimension of the Cartan subalgebra, $H$, is the same as that of $H_0^*$, the root space. This dimension, which is the same as the number of simple roots, is called the **rank** of the algebra. For a rank $n$ algebra, the Cartan matrix is the $n \times n$ matrix

$$A_{ij} = 2\frac{\langle \alpha_i, \alpha_j \rangle}{\langle \alpha_j, \alpha_j \rangle} \tag{VII.3}$$

where $\alpha_i, i = 1, \ldots n$ are the simple roots.

Clearly, the diagonal elements of the matrix are all equal to two. The matrix is not necessarily symmetric, but if $A_{ij} \neq 0$, then $A_{ji} \neq 0$. In fact, we have shown (see the discussion preceeding Eq. (VI.4) ) that the only possible values for the off-diagonal matrix elements are $0, \pm 1, \pm 2,$ and $\pm 3$. Indeed, since the scalar product of two different simple roots is non-positive, the off-diagonal elements can be only $0, -1, -2,$ and $-3$.

We have seen that $\langle \ , \ \rangle$ is a scalar product on the root space. The Schwarz inequality tells us that

$$\langle \alpha_i, \alpha_j \rangle^2 \leq \langle \alpha_i, \alpha_i \rangle \langle \alpha_j, \alpha_j \rangle \,, \tag{VII.4}$$

where the inequality is strict unless $\alpha_i$ and $\alpha_j$ are proportional. This cannot happen for $i \neq j$ since the simple roots are linearly independent. Thus we can write

$$A_{ij} A_{ji} < 4, \qquad i \neq j \,. \tag{VII.5}$$

It follows that if $A_{ij} = -2$ or $-3$, then $A_{ji} = -1$.

Consider again the SU(3) example. For simplicity, (and contrary to our choice above ) take the positive basis to be $\alpha_1$ and $\alpha_2$. Then since $\alpha_3 = \alpha_1 + \alpha_2$, the simple roots are also $\alpha_1$ and $\alpha_2$. We computed the relevant scalar products in Eq. (III.12):

$$\langle \alpha_1, \alpha_1 \rangle = \tfrac{1}{3}$$

$$\langle \alpha_1, \alpha_2 \rangle = - \tfrac{1}{6}$$

$$\langle \alpha_2, \alpha_2 \rangle = \tfrac{1}{3} \,. \tag{VII.6}$$

From this we compute the Cartan matrix

$$A = \begin{bmatrix} 2 & -1 \\ -1 & 2 \end{bmatrix} \,. \tag{VII.7}$$

The Cartan matrix, together with the ubiquitous Eq. (V.22), suffices to determine all the roots of a given simple Lie algebra. It is enough to determine all the positive roots, each of which can be written as a positive sum of simple roots: $\beta = \sum_i k_i \alpha_i$. We call $\sum_i k_i$ the **level** of the root $\beta$. Thus the simple roots are at the first level. Suppose we have determined all the roots up to the $n^{th}$level and wish to determine those at the level n+1. For each root $\beta$ at the $n^{th}$level, we must determine whether or not $\beta + \alpha_i$ is a root.

Since all the roots through the $n^{th}$level are known, it is known how far back the string of roots extends: $\beta, \beta - \alpha_i, \ldots \beta - m\alpha_i$. From this, we can compute

how far forward the string extends:$\beta, \beta + \alpha_i, \ldots \beta + p\alpha_i$. We just put our values
into Eq. (V.22):

$$m - p = 2\frac{\langle \beta, \alpha_i \rangle}{\langle \alpha_i, \alpha_i \rangle}$$

$$= \sum_j 2k_j \frac{\langle \alpha_j, \alpha_i \rangle}{\langle \alpha_i, \alpha_i \rangle}$$

$$= \sum_j k_j A_{ji} \ . \tag{VII.8}$$

In particular, $\beta + \alpha_i$ is a root if $p = m - \sum_j k_j A_{ji} > 0$ .

It is thus convenient to have an algorithm which keeps track of the $n$ quantities $\sum_j k_j A_{ji}$ for each root as it is determined. It is clear that this is accomplished
by adding to the $n$ quantities the $j^{th}$ row of the Cartan matrix whenever the $j^{th}$ simple
root is added to a root to form a new root.

Let us carry out this construction for SU(3). We begin by writing down the
Cartan matrix, then copying its rows to represent the simple roots:

$$\begin{bmatrix} 2 & -1 \\ -1 & 2 \end{bmatrix}$$

$$\boxed{2 \ -1} \qquad \boxed{-1 \ \ 2}$$

$$\boxed{1 \ \ 1}$$

Beginning with the root $\alpha_1$ we ask whether the addition of $\alpha_2$ produces a root in
the second level. (Remember that $2\alpha_1$ cannot be a root, nor can $\alpha_1 - \alpha_2$). Since
the second entry in the box for the first root is negative, the corresponding value of
$p$ in Eq. (VII.8) must be positive, so $\alpha_1 + \alpha_2$ is a root. The same conclusion would
be reached beginning with $\alpha_2$. Is there a root at level three? Looking back in the
$\alpha_1$ direction, $m = 1$. Since the first entry in the box for $\alpha_1 + \alpha_2$ is one, we have
$p = 0$ so we cannot add another $\alpha_1$. The same applies for $\alpha_2$. There are no roots
at the third level.

As a slightly more complex example, we display the result for the exceptional algebra $G_2$, which we shall discuss at greater length later:

$$\begin{bmatrix} 2 & -3 \\ -1 & 2 \end{bmatrix}$$

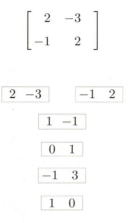

Not only does the Cartan matrix determine all of the roots, it determines the full commutation relations for the algebra. To see this, let us introduce the notation of Jacobson[1]. Start with any choice of normalization for $e_\alpha$ and $e_{-\alpha}$. We have shown that $[e_\alpha, e_{-\alpha}] = (e_\alpha, e_{-\alpha}) h_\alpha$. Now for every simple root, $\alpha_i$, define

$$e_i = e_{\alpha_i}$$

$$f_i = e_{-\alpha_i} \cdot 2 \left[ (e_{\alpha_i}, e_{-\alpha_i}) \langle \alpha_i, \alpha_i \rangle \right]^{-1}$$

$$h_i = h_{\alpha_i} \cdot \frac{2}{\langle \alpha_i, \alpha_i \rangle} \ . \tag{VII.9}$$

By direct computation we find

$$[e_i, f_j] = \delta_{ij} h_j$$

$$[h_i, e_j] = A_{ji} e_j$$

$$[h_i, f_j] = -A_{ji} f_j \ . \tag{VII.10}$$

The commutator $[e_i, f_j]$ vanishes unless $i = j$ since it would be proportional to $e_{\alpha_i - \alpha_j}$ and $\alpha_i - \alpha_j$ is not a root since $\alpha_i$ and $\alpha_j$ are simple.

A full basis for the Lie algebra can be obtained from the $e_i$'s, $f_i$'s and $h_i$'s. All of the raising operators can be written in the form $e_{i_1}, [e_{i_1}, e_{i_2}], [e_{i_1}, [e_{i_2}, e_{i_3}]]$, etc., and similarly for the lowering operators constructed from the $f$'s. Two elements obtained from commuting in this way the same set of $e$'s, but in different orders, are proportional with constant of proportionality being determined by the Cartan matrix through the commutation relations in Eq. (VII.10). Among the various orderings we choose one as a basis element. Following the same procedure for the $f$'s and adjoining the $h$'s we obtain a complete basis. The commutation relations among them can be shown to be determined by the simple commutation relations in Eq. (VII.10), that is, by the Cartan matrix.

The Cartan matrix thus contains all the information necessary to determine entirely the corresponding Lie algebra. Its contents can be summarized in an elegant diagrammatic form due to Dynkin. The **Dynkin diagram** of a semi-simple Lie algebra is constructed as follows. For every simple root, place a dot. As we shall show later, for a simple Lie algebra, the simple roots are at most of two sizes. Darken the dots corresponding to the smaller roots. Connect the $i^{th}$ and $j^{th}$ dots by a number of straight lines equal to $A_{ij} A_{ji}$. For a semi-simple algebra which is not simple, the diagram will have disjoint pieces, each of which corresponds to a simple algebra.

For SU(3) and $G_2$, we have the Cartan matrices and Dynkin diagrams shown below:

$$SU(3) = A_2 \qquad \begin{bmatrix} 2 & -1 \\ -1 & 2 \end{bmatrix}$$

$\alpha_1 \qquad \alpha_2$

$$G_2 \qquad \begin{bmatrix} 2 & -3 \\ -1 & 2 \end{bmatrix}$$

$$\alpha_1 \qquad \alpha_2$$

The darkened dot for $G_2$ corresponds to the second root, since the presence of the (-3) in the second row indicates that the second root is the smaller.

In subsequent sections we will determine the full set of Dynkin diagrams which represent simple Lie algebras. Here we anticipate the result somewhat in order to demonstrate how the Cartan matrix and Dynkin diagrams determine each other. Consider the Dynkin diagram:

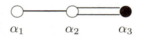

$$\alpha_1 \qquad \alpha_2 \qquad \alpha_3$$

The Cartan matrix is determined by noting that $A_{13} = A_{31} = 0$, since the first and third dots are not connected. Since one line connects the first and second points, we must have $A_{12} = A_{21} = -1$. The second and third points are connected by two lines so $A_{23}A_{32} = 2$. Since the third root is smaller than the second, it must be that $A_{23} = -2$ while $A_{32} = -1$. Thus we have

$$\begin{bmatrix} 2 & -1 & 0 \\ -1 & 2 & -2 \\ 0 & -1 & 2 \end{bmatrix}$$

## Footnote

1. JACOBSON, p. 121.

## References

Dynkin diagrams were first introduced in DYNKIN I. An excellent review of much of the material presented in this and other chapters is found in the Appendix to DYNKIN III.

## Exercises

1. Find the Dynkin diagram for

$$
\begin{bmatrix}
2 & -1 & 0 & 0 \\
-1 & 2 & -1 & 0 \\
0 & -2 & 2 & -1 \\
0 & 0 & -1 & 2
\end{bmatrix}.
$$

2. Find all the roots of $B_2$ whose Cartan matrix is

$$
\begin{bmatrix}
2 & -2 \\
-1 & 2
\end{bmatrix}.
$$

Draw a picture of the roots of $B_2$ like that in Fig. III.1.

3. Draw a picture of the roots of $G_2$ and compare with Fig. III.1.

# VIII. The Classical Lie Algebras

The general considerations of the previous chapter can be applied to the most familiar simple Lie algebras, the classical Lie algebras, SU(n), SO(n), and Sp(2n). These algebras are defined in terms of matrices and are simpler to visualize than some of the exceptional Lie algebras we shall encounter soon. The explicit construction of the Cartan subalgebra and the root vectors and roots for the classical algebras should make concrete our earlier results.

The space of all $n \times n$ matrices has a basis of elements $e_{ab}$ where the components of $e_{ab}$ are

$$(e_{ab})_{ij} = \delta_{ai}\delta_{bj} \; . \tag{VIII.1}$$

Thus the multiplication rule for the matrices is

$$e_{ab}e_{cd} = e_{ad}\delta_{bc} \tag{VIII.2}$$

and the commutator is

$$[e_{ab}, e_{cd}] = e_{ad}\delta_{bc} - e_{cb}\delta_{ad} . \tag{VIII.3}$$

The matrix $I = \sum_i e_{ii}$ commutes with all the basis elements. It thus forms the basis for a one-dimensional Abelian subalgebra. Consequently, the Lie algebra of all the $n \times n$ matrices is not semi-simple. However, if we restrict ourselves to traceless $n \times n$ matrices, we do obtain a semi-simple (in fact, simple) Lie algebra called $A_{n-1}$. This is the complex version of SU(n).

The elements of $A_{n-1}$ are linear combinations of the $e_{ab}$'s for $a \neq b$ and of elements $h = \sum_i \lambda_i e_{ii}$ where $\sum_i \lambda_i = 0$. From Eq. (VIII.3) we find the commutation relation

$$[h, e_{ab}] = (\lambda_a - \lambda_b)e_{ab} . \tag{VIII.4}$$

Thus $e_{ab}$ is a root vector corresponding to the root $\sum_i \lambda_i e_{ii} \rightarrow \lambda_a - \lambda_b$.

Let us choose as a basis for the root space

$$\alpha_1 : \qquad \sum_i \lambda_i e_{ii} \rightarrow \lambda_1 - \lambda_2$$

$$\alpha_2 : \qquad \sum_i \lambda_i e_{ii} \rightarrow \lambda_2 - \lambda_3$$

$$\cdots$$

$$\alpha_{n-1} : \qquad \sum_i \lambda_i e_{ii} \rightarrow \lambda_{n-1} - \lambda_n \tag{VIII.5}$$

and declare these positive with $\alpha_1 > \alpha_2 \ldots > \alpha_{n-1}$. It is easy to see that these same roots are the simple roots.

In order to find the scalar product $\langle \, , \, \rangle$, we first determine the Killing form as applied to elements of the Cartan algebra, using Eq. (IV.3) and Eq. (VIII.4):

$$\left(\sum_i \lambda_i e_{ii}, \sum_j \lambda'_j e_{jj}\right) = \operatorname{Tr} \operatorname{ad}\left(\sum_i \lambda_i e_{ii}\right) \operatorname{ad}\left(\sum_j \lambda'_j e_{jj}\right)$$

$$= \sum_{p,q} (\lambda_p - \lambda_q)(\lambda'_p - \lambda'_q)$$

$$= 2n \sum_p \lambda_p \lambda'_p . \tag{VIII.6}$$

The Killing form determines the connection between the Cartan subalgebra , $H$, and the root space $H_0^*$. That is, it enables us to find $h_{\alpha_i}$:

$$(h_{\alpha_j}, \sum_i \lambda_i e_{ii}) = \alpha_j\left(\sum_i \lambda_i e_{ii}\right)$$

$$= \lambda_j - \lambda_{j+1} . \tag{VIII.7}$$

Combining this with Eq. (VIII.6), we see that

$$h_{\alpha_i} = (e_{ii} - e_{i+1\ i+1})/(2n) \tag{VIII.8}$$

and

$$\langle \alpha_i, \alpha_j \rangle = (2\delta_{ij} - \delta_{i\ j+1} - \delta_{i+1\ j})/(2n) . \tag{VIII.9}$$

This agrees in particular with our earlier computation for SU(3). From the value of $\langle \alpha_i, \alpha_j \rangle$ we see that the Cartan matrix and Dynkin diagram are given by

$$A_n : \begin{bmatrix} 2 & -1 & 0 & & & & . & \\ -1 & 2 & -1 & & & & . & \\ 0 & -1 & . & & & & . & \\ & & & . & -1 & 0 & & \\ & & & -1 & 2 & -1 & & \\ . & . & . & 0 & -1 & 2 & \end{bmatrix}$$

$$\underset{\alpha_1}{\circ} \!\!-\!\!-\!\!-\!\! \underset{\alpha_2}{\circ} \!\!-\!\!-\!\! \cdots \!\!-\!\!-\!\! \underset{\alpha_n}{\circ}$$

where we have chosen to represent $A_n$ rather than $A_{n-1}$.

We next consider the **symplectic group** $\mathrm{Sp}(2m)$ and its associated Lie algebra. The group consists of the $2m \times 2m$ matrices $A$ with the property $A^t J A = J$ where $(\ )^t$ indicates transpose and $J$ is the $2m \times 2m$ matrix

$$
J = \begin{bmatrix} 0 & I \\ -I & 0 \end{bmatrix} .
\tag{VIII.10}
$$

The corresponding requirement for the Lie algebra is obtained by writing $A = \exp(\mathcal{A}) \approx I + \mathcal{A}$. Thus we have $\mathcal{A}^t = J\mathcal{A}J$. In terms of $m \times m$ matrices, we can write

$$
\mathcal{A} = \begin{bmatrix} \mathcal{A}_1 & \mathcal{A}_2 \\ \mathcal{A}_3 & \mathcal{A}_4 \end{bmatrix}
\tag{VIII.11}
$$

and find the restrictions $\mathcal{A}_1^t = -\mathcal{A}_4, \mathcal{A}_2 = \mathcal{A}_2^t, \mathcal{A}_3 = \mathcal{A}_3^t$. In accordance with these, we choose the following basis elements $(j, k \leq m)$:

$$
\begin{aligned}
e_{jk}^1 &= e_{jk} - e_{k+m,j+m} , \\
e_{jk}^2 &= e_{j,k+m} + e_{k,j+m} , \qquad j \leq k \\
e_{jk}^3 &= e_{j+m,k} + e_{k+m,j} , \qquad j \leq k .
\end{aligned}
\tag{VIII.12}
$$

The Cartan subalgebra has a basis $h_j = e_{jj}^1$. By direct computation we find that if $h = \sum_i h_i \lambda_i$,

$$
\begin{aligned}
\left[h, e_{jk}^1\right] &= +(\lambda_j - \lambda_k)e_{jk}^1 , \qquad j \neq k \\
\left[h, e_{jk}^2\right] &= +(\lambda_j + \lambda_k)e_{jk}^2 , \qquad j \leq k \\
\left[h, e_{jk}^3\right] &= -(\lambda_j + \lambda_k)e_{jk}^3 , \qquad j \leq k .
\end{aligned}
\tag{VIII.13}
$$

We take as an ordered basis of roots $\alpha_1(h) = \lambda_1 - \lambda_2$, $\alpha_2(h) = \lambda_2 - \lambda_3$, $\dots \alpha_{m-1}(h) = \lambda_{m-1} - \lambda_m$, $\alpha_m(h) = 2\lambda_m$. With this ordering, the $\alpha_i$'s are themselves simple roots. For example, the root $\alpha(h) = \lambda_{m-1} + \lambda_m$ is not simple since it is the sum of $\alpha_{m-1}$ and $\alpha_m$.

We calculate the Killing form on the Cartan subalgebra explicitly by considering in turn the contribution of each root to the trace which defines the form.

$$\left( \sum_i \lambda_i h_i, \sum_j \lambda'_j h_j \right) = \sum_{p,q} (\lambda_p - \lambda_q)(\lambda'_p - \lambda'_q) + 2 \sum_{p \leq q} (\lambda_p + \lambda_q)(\lambda'_p + \lambda'_q)$$

$$= \sum_{p,q} [(\lambda_p - \lambda_q)(\lambda'_p - \lambda'_q) + (\lambda_p + \lambda_q)(\lambda'_p + \lambda'_q)] + \sum_p 4\lambda_p \lambda'_p$$

$$= 4(m+1) \sum_p \lambda_p \lambda'_p . \qquad (VIII.14)$$

We easily see then that

$$h_{\alpha_i} = \frac{(h_i - h_{i+1})}{4(m+1)} , \qquad i < m$$

$$h_{\alpha_m} = \frac{h_m}{2(m+1)} . \qquad (VIII.15)$$

Since $(h_i, h_j) = \delta_{ij} 4(m+1)$, we can compute directly all the terms we need for the Cartan matrix:

$$\langle \alpha_i, \alpha_j \rangle = \frac{1}{4(m+1)} (2\delta_{ij} - \delta_{i\ j+1} - \delta_{i+1\ j}), \qquad i, j \neq m$$

$$\langle \alpha_i, \alpha_m \rangle = -\frac{1}{2(m+1)} \delta_{i+1\ m}, \qquad i \neq m$$

$$\langle \alpha_m, \alpha_m \rangle = \frac{1}{(m+1)} . \qquad (VIII.16)$$

The Lie algebra which is associated with Sp(2n) is denoted $C_n$. From Eq. (VIII.16) we derive its Cartan matrix and Dynkin diagram:

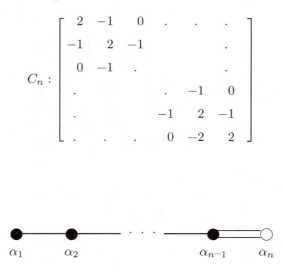

The **orthogonal groups** are given by matrices which satisfy $A^t A = I$. Using the correspondence between elements of the group and elements of the Lie algebra as discussed in Chapter I, $A = \exp \mathcal{A} \approx I + \mathcal{A}$, we see that the requirement is $\mathcal{A} + \mathcal{A}^t = 0$. Clearly these matrices have only off-diagonal elements. As a result, it would be hard to find the Cartan subalgebra as we did for $A_n$ and $C_n$ by using the diagonal matrices. To avoid this problem, we perform a unitary transformation on the matrices $A$. This will yield an equivalent group of matrices obeying a modified condition. Let us write

$$A = UBU^\dagger , \qquad\qquad (VIII.17)$$

so that

$$A^t A = U^{\dagger t} B^t U^t U B U^\dagger = I. \qquad\qquad (VIII.18)$$

Setting $K = U^t U$, we have $B^t K B = K$. Writing $B \approx I + \mathcal{B}$, we have

$$\mathcal{B}^t K + K \mathcal{B} = 0. \qquad\qquad (VIII.19)$$

A convenient choice for the even dimensional case, $n = 2m$, is

$$U = \frac{1}{\sqrt{2}} \begin{bmatrix} i & -i \\ -1 & -1 \end{bmatrix} , \tag{VIII.20}$$

so that

$$K = \begin{bmatrix} 0 & I \\ I & 0 \end{bmatrix} . \tag{VIII.21}$$

Representing $\mathcal{B}$ in terms of $m \times m$ matrices,

$$\mathcal{B} = \begin{bmatrix} \mathcal{B}_1 & \mathcal{B}_2 \\ \mathcal{B}_3 & \mathcal{B}_4 \end{bmatrix} \tag{VIII.22}$$

the condition becomes

$$\mathcal{B}_1 = -\mathcal{B}_4^t , \quad \mathcal{B}_2 = -\mathcal{B}_2^t , \quad \mathcal{B}_3 = -\mathcal{B}_3^t . \tag{VIII.23}$$

We can now select a basis of matrices obeying these conditions:

$$e_{jk}^1 = e_{j,k} - e_{k+m,j+m},$$

$$e_{jk}^2 = e_{j,k+m} - e_{k,j+m}, \qquad j < k$$

$$e_{jk}^3 = e_{j+m,k} - e_{k+m,j}, \qquad j < k \tag{VIII.24}$$

and designate the basis for the Cartan subalgebra by

$$h_j = e_{jj}^1 . \tag{VIII.25}$$

Writing a general element of the Cartan subalgebra as

$$h = \sum_i \lambda_i h_i , \tag{VIII.26}$$

we compute the various roots

$$[h, e^1_{jk}] = (\lambda_j - \lambda_k)e^1_{jk} \qquad j \neq k$$
$$[h, e^2_{jk}] = (\lambda_j + \lambda_k)e^2_{jk} \qquad j < k$$
$$[h, e^3_{jk}] = -(\lambda_j + \lambda_k)e^3_{jk} \qquad j < k .$$ 
$$\text{(VIII.27)}$$

Note that for $e^2_{jk}$ and $e^3_{jk}$ we must have $j \neq k$ or else the matrix vanishes. Thus there are no roots corresponding to $\pm 2\lambda_j$. We may take as a basis of simple roots $\alpha_1(h) = \lambda_1 - \lambda_2, \alpha_2(h) = \lambda_2 - \lambda_3, \ldots \alpha_{m-1}(h) = \lambda_{m-1} - \lambda_m, \alpha_m(h) = \lambda_{m-1} + \lambda_m$.

The Killing form restricted to the Cartan subalgebra is given by

$$(\sum_i \lambda_i h_i, \sum_j \lambda'_j h_j) = \sum_{i \neq j}(\lambda_i - \lambda_j)(\lambda'_i - \lambda'_j) + 2\sum_{i<j}(\lambda_i + \lambda_j)(\lambda'_i + \lambda'_j)$$

$$= \sum_{i,j}[(\lambda_i - \lambda_j)(\lambda'_i - \lambda'_j) + (\lambda_i + \lambda_j)(\lambda'_i + \lambda'_j)] - \sum_i 4\lambda_i \lambda'_i$$

$$= 4(m-1)\sum_i \lambda_i \lambda'_i .$$ 
$$\text{(VIII.28)}$$

From this relation we can determine the $h_{\alpha_i}$'s:

$$h_{\alpha_i} = \frac{h_i - h_{i+1}}{4(m-1)} , \qquad i < m \qquad\qquad \text{(VIII.29}a\text{)}$$

$$h_{\alpha_m} = \frac{h_{m-1} + h_m}{4(m-1)} . \qquad\qquad \text{(VIII.29}b\text{)}$$

The scalar products of the roots are now easily computed:

$$\langle \alpha_i, \alpha_j \rangle = [2\delta_{ij} - \delta_{ij+1} - \delta_{i+1j}]/[4(m-1)] \qquad i, j < m$$

$$\langle \alpha_m, \alpha_m \rangle = 1/[2(m-1)]$$

$$\langle \alpha_{m-1}, \alpha_m \rangle = 0,$$

$$\langle \alpha_{m-2}, \alpha_m \rangle = -1/[4(m-1)] . \qquad\qquad \text{(VIII.30)}$$

Thus the Cartan matrix and Dynkin diagram are

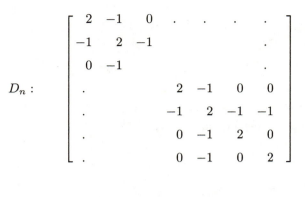

$$D_n : \begin{bmatrix} 2 & -1 & 0 & \cdot & \cdot & \cdot & \cdot \\ -1 & 2 & -1 & & & & \cdot \\ 0 & -1 & & & & & \cdot \\ \cdot & & & 2 & -1 & 0 & 0 \\ \cdot & & & -1 & 2 & -1 & -1 \\ \cdot & & & 0 & -1 & 2 & 0 \\ \cdot & & & 0 & -1 & 0 & 2 \end{bmatrix}$$

$F$

or the odd dimensional case of the orthogonal group, we proceed the same way except that we set

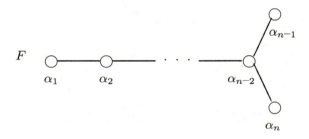

$$U = \tfrac{1}{\sqrt{2}} \begin{bmatrix} \sqrt{2} & 0 & 0 \\ 0 & i_m & -i_m \\ 0 & -1_m & -1_m \end{bmatrix} \tag{VIII.31}$$

so that

$$K = \begin{bmatrix} 1 & 0 & 0 \\ 0 & 0_m & 1_m \\ 0 & 1_m & 0_m \end{bmatrix} \tag{VIII.32}$$

where the subscript $m$ indicates an $m \times m$ matrix. The corresponding matrix $\mathcal{B}$ may be parameterized as

$$B = \begin{bmatrix} b_1 & c_1 & c_2 \\ d_1 & \mathcal{B}_1 & \mathcal{B}_2 \\ d_2 & \mathcal{B}_3 & \mathcal{B}_4 \end{bmatrix} . \qquad \text{(VIII.33)}$$

For the $2m \times 2m$ pieces of the matrix, the conditions are the same as for the even dimensional orthogonal algebra. The constraints on the new matrices are

$$b_1 = 0 , \qquad c_1 = -d_2^t , \qquad c_2 = -d_1^t . \qquad \text{(VIII.34)}$$

Thus we must add to our basis for the $2m$ dimensional orthogonal algebra the elements $(1 \le j \le m)$ :

$$e_j^4 = e_{0j} - e_{j+m\ 0} ; \qquad e_j^5 = e_{j0} - e_{0\ j+m} . \qquad \text{(VIII.35)}$$

The corresponding roots are seen to be

$$[h, e_j^4] = -\lambda_j e_j^4 ; \qquad [h, e_j^5] = \lambda_j e_j^5 . \qquad \text{(VIII.36)}$$

Using these new roots, together with those found for the even dimensional case, we compute the Killing form

$$\left( \sum_i \lambda_i h_i, \sum_j \lambda_j' h_j \right)$$

$$= \sum_{i \ne j} (\lambda_i - \lambda_j)(\lambda_i' - \lambda_j') + 2 \sum_{i<j} (\lambda_i + \lambda_j)(\lambda_i' + \lambda_j') + 2 \sum_i \lambda_i \lambda_i'$$

$$= 4(m - \tfrac{1}{2}) \sum_i \lambda_i \lambda_i' . \qquad \text{(VIII.37)}$$

From this we can infer the values

$$h_{\alpha_i} = \frac{h_{\alpha_i} - h_{\alpha_{i+1}}}{4(m - \frac{1}{2})} , \qquad i < m$$

$$h_{\alpha_m} = \frac{h_{\alpha_m}}{4(m - \frac{1}{2})} \qquad\qquad \text{(VIII.38)}$$

where now the simple roots have the values $\alpha_1(h) = \lambda_1 - \lambda_2, \alpha_2(h) = \lambda_2 - \lambda_3, \ldots \alpha_{m-1}(h) = \lambda_{m-1} - \lambda_m, \alpha_m(h) = \lambda_m$. Note that the last of these was not even a root for the even dimensional case. Using the Killing form, it is easy to compute the scalar product on the root space:

$$\langle \alpha_i, \alpha_j \rangle = \frac{1}{4(m - \frac{1}{2})}(2\delta_{ij} - \delta_{i\ j+1} - \delta_{i+1\ j}), \qquad i < m$$

$$\langle \alpha_m, \alpha_i \rangle = 0, \qquad i < m - 1$$

$$\langle \alpha_m, \alpha_{m-1} \rangle = -\frac{1}{4(m - \frac{1}{2})} ,$$

$$\langle \alpha_m, \alpha_m \rangle = \frac{1}{4(m - \frac{1}{2})} . \qquad\qquad \text{(VIII.39)}$$

Accordingly, the Cartan matrix and Dynkin diagram are

$$B_n : \begin{bmatrix} 2 & -1 & 0 & . & . & & . \\ -1 & 2 & -1 & & & & . \\ 0 & -1 & & & & & . \\ . & & & 2 & -1 & 0 \\ . & & & -1 & 2 & -2 \\ . & . & . & 0 & -1 & 2 \end{bmatrix}$$

$$\alpha_1 \qquad \alpha_2 \qquad\qquad \alpha_{n-1} \qquad \alpha_n$$

Notice the similarity between $B_n$ and $C_n$. In the Cartan matrix they differ only by the interchange of the last off-diagonal elements. The corresponding change in the Dynkin diagrams is to reverse the shading of the dots.

# References

This material is discussed in DYNKIN I, JACOBSON, pp. 135-141, and MILLER, pp. 351–354.

# Exercise

1. Starting with the Dynkin diagrams, construct drawings of the roots of $B_2$, $D_2$, $A_3$, $B_3$, and $C_3$.

# IX. The Exceptional Lie Algebras

We have displayed the four series of classical Lie algebras and their Dynkin diagrams. How many more simple Lie algebras are there? Surprisingly, there are only five. We may prove this by considering a set of vectors (candidates for simple roots) $\gamma_i \subset H_0^*$ and defining a matrix (analogous to the Cartan matrix)[1]:

$$M_{ij} = 2\frac{\langle \gamma_i, \gamma_j \rangle}{\langle \gamma_j, \gamma_j \rangle} \qquad \text{(IX.1)}$$

and an associated diagram (analogous to the Dynkin diagram), where the i [th] and j [th] points are joined by $M_{ij}M_{ji}$ lines. The set $\gamma_i$ is called **allowable** , (in Jacobson's usage) if

  *i.* The $\gamma_i$ are linearly independent, that is, if det M $\neq 0$.

  *ii.* $M_{ij} \leq 0$ for $i \neq j$.

  *iii.* $M_{ij}M_{ji} = 0, 1, 2,$ or 3.

With these definitions, we can prove a series of lemmas:

1. Any subset of an allowable set is allowable. *Proof:* Since a subset of a linearly independent set is linearly independent, (i) is easy. Equally obvious are (ii) and (iii).

2. An allowable set has more points than joined pairs. *Proof:* Let $\gamma = \sum_i \gamma_i \langle \gamma_i, \gamma_i \rangle^{-\frac{1}{2}}$. Since the set is linearly independent, $\gamma \neq 0$ so $\langle \gamma, \gamma \rangle > 0$. Thus

$$0 < \langle \gamma, \gamma \rangle = \sum_{i<j} 2 \frac{\langle \gamma_i, \gamma_j \rangle}{\langle \gamma_i, \gamma_i \rangle^{\frac{1}{2}} \langle \gamma_j, \gamma_j \rangle^{\frac{1}{2}}} + \text{no. of points}$$

$$0 < -\sum_{i<j} [M_{ij} M_{ji}]^{\frac{1}{2}} + \text{no. of points} . \tag{IX.2}$$

For each pair of joined points, $M_{ij} M_{ji}$ is at least unity, so

$$\text{no. of joined pairs} < \text{no. of points.}$$

3. An allowable set's diagram has no loops. *Proof:* If it did, there would be a subset with at least as many joined pairs as points.

4. If an allowable set has a diagram with a chain of points joined only to successive points by single lines, there is an allowable set whose diagram is the same except that the chain is shrunk to a point. *Proof:* Let the chain be $\beta_1, \beta_2, \ldots \beta_m$ and let $\beta = \sum_i \beta_i$. Now

$$\langle \beta, \beta \rangle = \sum_i \langle \beta_i, \beta_i \rangle + 2 \sum_{i<j} \langle \beta_i, \beta_j \rangle$$

$$= m \langle \beta_1, \beta_1 \rangle - (m-1) \langle \beta_1, \beta_1 \rangle$$

$$= \langle \beta_1, \beta_1 \rangle \tag{IX.3}$$

so $\beta$ is the same size as the individual points in the chain. Moreover, if $\gamma$ is joined to the chain at the end, say to $\beta_1$, then $\langle \gamma, \beta_1 \rangle = \langle \gamma, \beta \rangle$, since $\langle \gamma, \beta_j \rangle = 0$ for all $j \neq 1$.

5. No more than three lines emanate from a vertex of an allowable diagram. *Proof:* Suppose $\gamma_1, \gamma_2, \ldots \gamma_m$ are connected to $\gamma_0$. Then $\langle \gamma_i, \gamma_j \rangle = 0$, for $i, j \neq 0$ since there are no loops. Since $\gamma_0$ is linearly independent of the $\gamma_i$, its magnitude squared is greater than the sum of the squares of its components along the orthogonal directions $\gamma_i \langle \gamma_i, \gamma_i \rangle^{-\frac{1}{2}}$:

$$\langle \gamma_0, \gamma_0 \rangle > \sum_i \langle \gamma_0, \gamma_i \rangle^2 \langle \gamma_i, \gamma_i \rangle^{-1} . \tag{IX.4}$$

Thus $4 > \sum_i M_{0i} M_{i0}$. But $M_{0i} M_{i0}$ is the number of segments joining $\gamma_0$ and $\gamma_i$.

6. The only allowable configuration with a triple line is

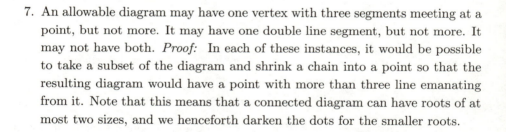

7. An allowable diagram may have one vertex with three segments meeting at a point, but not more. It may have one double line segment, but not more. It may not have both. *Proof:* In each of these instances, it would be possible to take a subset of the diagram and shrink a chain into a point so that the resulting diagram would have a point with more than three line emanating from it. Note that this means that a connected diagram can have roots of at most two sizes, and we henceforth darken the dots for the smaller roots.

8. The diagrams

and

are not allowable. *Proof:* Consider the determinant of $M$ for the first diagram:

$$\begin{bmatrix} 2 & -1 & 0 & 0 & 0 \\ -1 & 2 & -1 & 0 & 0 \\ 0 & -2 & 2 & -1 & 0 \\ 0 & 0 & -1 & 2 & -1 \\ 0 & 0 & 0 & -1 & 2 \end{bmatrix}.$$

We see that if we add the first and last columns, plus twice the second and fourth, plus three times the third, we get all zeros. Thus the determinant vanishes. The matrix for the second diagram is just the transpose of the first.

9. The only diagrams with a double line segment which may be allowable are of the form:

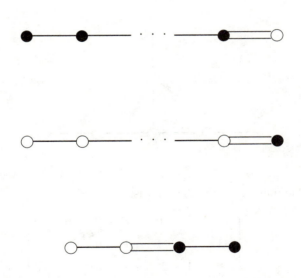

10. By (7) above, the only diagrams with a branch in them are of the form:

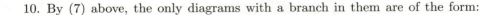

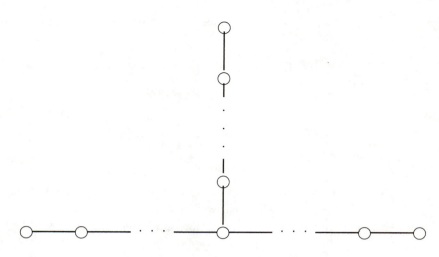

11. The diagram below is not allowable

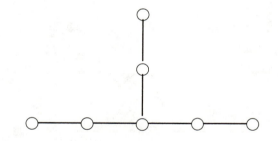

*Proof:* The matrix for the diagram is:

$$\begin{bmatrix} 2 & -1 & 0 & 0 & 0 & 0 & 0 \\ -1 & 2 & -1 & 0 & 0 & 0 & 0 \\ 0 & -1 & 2 & -1 & 0 & -1 & 0 \\ 0 & 0 & -1 & 2 & -1 & 0 & 0 \\ 0 & 0 & 0 & -1 & 2 & 0 & 0 \\ 0 & 0 & -1 & 0 & 0 & 2 & -1 \\ 0 & 0 & 0 & 0 & 0 & -1 & 2 \end{bmatrix}$$

Straightforward manipulation like that above shows that the determinant vanishes.

12. The only allowable diagrams with a branch in them are of the form:

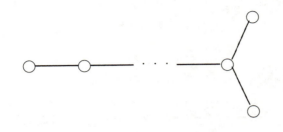

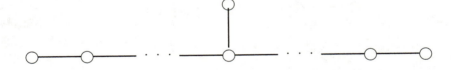

13. The diagram below is not allowable. This is proved simply by evaluating the associated determinant and showing it vanishes.

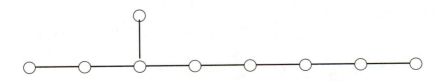

14. The complete list of allowable configurations is

$A_n$

$B_n$

$C_n$

$D_n$

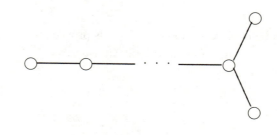

$G_2$

$F_4$

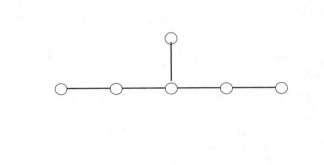

$E_6$

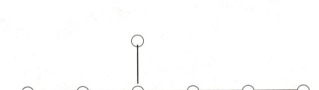

$E_7$

$E_8$

Above are given the names use to designate the five **exceptional Lie algebras**. So far we have only excluded all other possibilities. In fact, these five diagrams do correspond to simple Lie algebras.

## Footnote

1. Throughout the chapter we follow the approach of JACOBSON, pp. 128–135.

## Exercise

1. Prove #11 and #13 above.

# X. More on Representations

We continue our discussion of Section IV. As before, a representation is a mapping of the elements of the Lie algebra into linear operators,

$$e_\alpha \to E_\alpha$$

$$h_i \to H_i \tag{X.1}$$

which preserves the commutation relations of the Lie algebra. The operators $E$ and $H$ act on a vector space with elements generically denoted $\phi$. We can select a basis in which the $H$'s are diagonal.[1] Thus we can write

$$H\phi^M = M(h)\phi^M \tag{X.2}$$

where $M \in H_0^*$ is called a **weight** and $\phi^M$ is a **weight vector**.

The weights come in sequences with successive weights differing by roots of the Lie algebra. We have seen that if a complete string of roots is $M + p\alpha, \ldots M, \ldots M - m\alpha$, then (see Eq. (V.22))

$$m - p = 2\frac{\langle M, \alpha \rangle}{\langle \alpha, \alpha \rangle} \, . \tag{X.3}$$

A finite dimensional irreducible representation must have a highest weight, that is, a weight $\Lambda$ such that every other weight is less than $\Lambda$, where the ordering is determined in the usual fashion (That is, we pick a basis of roots and order it. A weight is positive if, when expanded in this ordered basis, the first non-zero coefficient is positive, and we say $M_1 > M_2$ if $M_1 - M_2 > 0$.)

Let $\{\alpha_i\}$ be a basis of simple roots and let $\Lambda$ be the highest weight of an irreducible representation. Then $\Lambda + \alpha_i$ is not a weight. Thus by Eq. (X.3),

$$\Lambda_i = 2\frac{\langle \Lambda, \alpha_i \rangle}{\langle \alpha_i, \alpha_i \rangle} \geq 0 \, . \tag{X.4}$$

Each greatest weight, $\Lambda$, is thus specified by a set of non-negative integers called **Dynkin coefficients**:

$$\Lambda_i = 2\frac{\langle \Lambda, \alpha_i \rangle}{\langle \alpha_i, \alpha_i \rangle} \, . \tag{X.5}$$

We could use the inverse of the Cartan matrix to determine the precise expansion of $\Lambda$ in terms of the simple roots, but this is rarely worthwhile.

Given the Dynkin coefficients of the highest weight, it is easy to determine the full set of weights in the irreducible representation, expressed again in terms of their Dynkin coefficients. The algorithm is similar to that we used to find all the roots of a Lie algebra from its Cartan matrix. Given a weight, $M$, we need to determine whether $M - \alpha_j$ is also a weight. Since we begin with the highest weight and work down, we know the value of $p$ in Eq. (X.3). We shall keep track of the integers

$$M_i = 2\frac{\langle M, \alpha_i \rangle}{\langle \alpha_i, \alpha_i \rangle} \, . \tag{X.6}$$

If
$$m_j = p_j + M_j > 0 , \tag{X.7}$$

then we know we can subtract the root $\alpha_j$ from the root $M$ to obtain another root. We record the Dynkin coefficients of $M - \alpha_j$ by subtracting from $M_i$ the quantities $A_{ji}$. This is most easily carried out by writing the Cartan matrix at the top of the computation.

Consider an example for SU(3) (or $A_2$ in the other notation). Let us determine the weights corresponding to the irreducible representation whose highest weight has Dynkin coefficients (1,0).

$$\begin{bmatrix} 2 & -1 \\ -1 & 2 \end{bmatrix}$$

$$\boxed{1 \quad 0}$$

$$\boxed{-1 \quad 1}$$

$$\boxed{0 \quad -1}$$

The Dynkin coefficients are entered in the boxes and successive rows are obtained by subtracting the appropriate row of the Cartan matrix. It is easy to see that the highest weight here can be expanded in simple roots as

$$\Lambda = \tfrac{2}{3}\alpha_1 + \tfrac{1}{3}\alpha_2 . \tag{X.8}$$

Thus the weights of this three dimensional representation are

$$\tfrac{2}{3}\alpha_1 + \tfrac{1}{3}\alpha_2 ; \quad -\tfrac{1}{3}\alpha_1 + \tfrac{1}{3}\alpha_2 ; \quad -\tfrac{1}{3}\alpha_1 - \tfrac{2}{3}\alpha_2 .$$

Of course, if we had started with Dynkin coefficients (0,1), we would have obtained a three dimensional representation with weights

$$\tfrac{1}{3}\alpha_1 + \tfrac{2}{3}\alpha_2 \ ; \quad \tfrac{1}{3}\alpha_1 - \tfrac{1}{3}\alpha_2 \ ; \quad -\tfrac{2}{3}\alpha_1 - \tfrac{1}{3}\alpha_2 \ .$$

Actually, we have relied on our previous knowledge of SU(3) to assert that these representations are three dimensional. All we have seen is that there are three different weights. It is often the case that a weight may correspond to more than one (linearly independent) weight vector, so that the **weight space** may be more than one dimensional. Consider for example the SU(3) representation with Dynkin coefficients (1,1), the familiar adjoint representation:

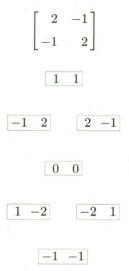

$$\begin{bmatrix} 2 & -1 \\ -1 & 2 \end{bmatrix}$$

This representation is eight dimensional. The weight with Dynkin coefficients (0,0) corresponds to a two dimensional space. Indeed, since this is the adjoint representation, we recognize that this space coincides with the Cartan subalgebra. The procedure for determining the dimensionality of a weight space will be discussed later.

As two additional examples, consider the representations of SO(10) ($D_5$) specified by the Dynkin coefficients (1,0,0,0,0) and (0,0,0,0,1) where the simple roots are numbered:

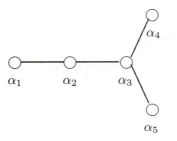

We have then the schemes:

$$\begin{bmatrix} 2 & -1 & 0 & 0 & 0 \\ -1 & 2 & -1 & 0 & 0 \\ 0 & -1 & 2 & -1 & -1 \\ 0 & 0 & -1 & 2 & 0 \\ 0 & 0 & -1 & 0 & 2 \end{bmatrix}$$

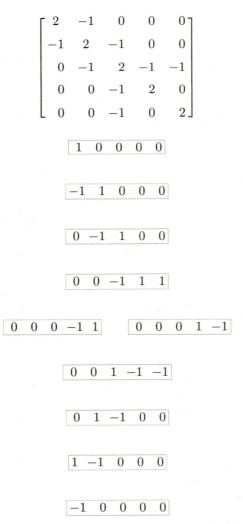

and

$$\begin{bmatrix} 2 & -1 & 0 & 0 & 0 \\ -1 & 2 & -1 & 0 & 0 \\ 0 & -1 & 2 & -1 & -1 \\ 0 & 0 & -1 & 2 & 0 \\ 0 & 0 & -1 & 0 & 2 \end{bmatrix}$$

$$\boxed{0 \quad 0 \quad 0 \quad 0 \quad 1}$$

$$\boxed{0 \quad 0 \quad 1 \quad 0 \quad -1}$$

$$\boxed{0 \quad 1 \quad -1 \quad 1 \quad 0}$$

$$\boxed{1 \quad -1 \quad 0 \quad 1 \quad 0} \qquad \boxed{0 \quad 1 \quad 0 \quad -1 \quad 0}$$

$$\boxed{-1 \quad 0 \quad 0 \quad 1 \quad 0} \qquad \boxed{1 \quad -1 \quad 1 \quad -1 \quad 0}$$

$$\boxed{-1 \quad 0 \quad 1 \quad -1 \quad 0} \qquad \boxed{1 \quad 0 \quad -1 \quad 0 \quad 1}$$

$$\boxed{-1 \quad 1 \quad -1 \quad 0 \quad 1} \qquad \boxed{1 \quad 0 \quad 0 \quad 0 \quad -1}$$

$$\boxed{0 \quad -1 \quad 0 \quad 0 \quad 1} \qquad \boxed{-1 \quad 1 \quad 0 \quad 0 \quad -1}$$

$$\boxed{0 \quad -1 \quad 1 \quad 0 \quad -1}$$

$$\boxed{0 \quad 0 \quad -1 \quad 1 \quad 0}$$

$$\boxed{0 \quad 0 \quad 0 \quad -1 \quad 0}$$

New representations may be obtained by taking products of representations. This procedure when applied to SU(2) is the familiar addition of angular momentum in quantum mechanics. Suppose we have two representations, $x \to X^{(1)}$ and $x \to X^{(2)}$ where $x \in L$ and $X^{(1)}$ and $X^{(2)}$ are linear operators on vector spaces whose basis elements will be denoted by $\phi$ and $\eta$ respectively:

$$X^{(1)}\phi_i = \sum_j X_{ij}^{(1)}\phi_j \tag{X.9a}$$

$$X^{(2)}\eta_i = \sum_j X_{ij}^{(2)}\eta_j \ . \tag{X.9b}$$

Here $X_{ij}^{(1)}$ and $X_{ij}^{(2)}$ are coefficients, not operators. We can define a product representation on the product space whose basis elements are of the form $\phi_i \otimes \eta_j$ as follows:

$$X\phi_i \otimes \eta_j = \sum_k X_{ik}^{(1)}\phi_k \otimes \eta_j + \sum_l \phi_i \otimes X_{jl}^{(2)}\eta_l \ . \tag{X.10}$$

For the rotation group, we might write $J = L + S$ and $\phi$ and $\eta$ might represent the spatial and spin parts of the wave function.

If $x$ is an element of the Cartan subalgebra we indicate it by $h$ and its representation by $H$. If $\phi$ and $\eta$ are weight vectors with weights $M^{(1)}$ and $M^{(2)}$, then $\phi \otimes \eta$ is a weight vector of the product representation with weight $M^{(1)} + M^{(2)}$, as we see from Eq. (X.10). If the highest weights of the two representations are $\Lambda^{(1)}$ and $\Lambda^{(2)}$, then the highest weight of the product representation is $\Lambda^{(1)} + \Lambda^{(2)}$.

Our construction of the weights of an irreducible representation from the Dynkin coefficients of its highest weight shows that all the weights are determined by the highest weight. It is also possible to show that the weight space of the highest weight is always one-dimensional for an irreducible representation. Thus each product of irreducible representations contains one irreducible representation whose highest weight is the sum of the highest weights of the two representations forming it.

As an example, consider again SU(3). The three dimensional representation may be represented by (1,0) and the other three dimensional representation by

(0,1). Their product must contain the representation (1,1), which is in fact the eight dimensional representation.

Consider the special case in which the representations being multiplied are identical. The product space can be broken into two parts, a symmetric part with basis $\phi_i \otimes \phi_j + \phi_j \otimes \phi_i$, and an anti-symmetric part, with basis $\phi_i \otimes \phi_j - \phi_j \otimes \phi_i$. If the highest weight of the representation carried by $\phi$ is $\Lambda$, then the highest weight carried by the symmetric space is $2\Lambda$. The anti-symmetric space does not contain the vector with this weight since it is symmetric . The highest weight in the anti-symmetric space is found by taking the sum of the highest and the next-to-highest weights.

Again, a simple example may be taken from SU(3). Consider $3 \times 3$ (i.e. $(1,0) \times (1,0)$). The second highest weight in (1,0) is (-1,1). Thus the anti-symmetric space carries the representation whose highest weight is (1,0)+(-1,1)=(0,1). This is the $3^*$. The symmetric space carries the (2,0), the 6 of SU(3). In general, however, the product contains more than two irreducible components.

It is possible to extend the anti-symmetrization procedure by taking the n-fold anti-symmetric product of a given representation. It is clear that the three fold anti-symmetric product will contain a representation whose highest weight is the sum of the three highest weights of the irreducible representation from which it is made, and so on. Similarly, the n-fold symmetric product will contain an irreducible representation whose highest weight is n-times the highest weight of the initial irreducible representation.

These procedures are especially easy to apply to $A_n$, beginning with the **fundamental representation**,$(1,0,\ldots)$. Calculating the weights of this representation, we quickly see that the two-fold anti-symmetrization yields a highest weight $(0,1,0,\ldots)$, the three-fold anti-symmetrization $(0,0,1,0,\ldots)$, and so on.

In fact, combining these operations we can produce any of the irreducible representations of $A_n$. To produce the representation with highest weight $(m_1, m_2, m_3, \ldots)$, we take the $m_1$-fold symmetric product of $(1,0,0,\ldots)$ and the $m_2$-fold symmetric product of $(0,1,0,\ldots)$ and form their product. The irreducible representation with highest weight in the product is $(m_1, m_2, 0, \ldots)$. We continue in this fashion to build $(m_1, m_2, m_3, \ldots)$.

The representations with Dynkin coefficients all equal to zero except for one entry of unity are called **basic representations**. It is clear that every representation can be formed from basic representations simply using the highest weights of product representations. Moreover, for $A_n$, we have seen that every basic representation can be obtained from a single fundamental representation.

Consider, on the other hand, $B_3$, $(O(7))$. We display the weights of the representations $(1,0,0)$ and $(0,0,1)$.

$$\begin{bmatrix} 2 & -1 & 0 \\ -1 & 2 & -2 \\ 0 & -1 & 2 \end{bmatrix}$$

$$\boxed{\begin{array}{ccc} 1 & 0 & 0 \end{array}}$$

$$\boxed{\begin{array}{ccc} -1 & 1 & 0 \end{array}}$$

$$\boxed{\begin{array}{ccc} 0 & -1 & 2 \end{array}}$$

$$\boxed{\begin{array}{ccc} 0 & 0 & 0 \end{array}}$$

$$\boxed{\begin{array}{ccc} 0 & 1 & -2 \end{array}}$$

$$\boxed{\begin{array}{ccc} 1 & -1 & 0 \end{array}}$$

$$\boxed{\begin{array}{ccc} -1 & 0 & 0 \end{array}}$$

$$\begin{bmatrix} 2 & -1 & 0 \\ -1 & 2 & -2 \\ 0 & -1 & 2 \end{bmatrix}$$

$$\boxed{0 \quad 0 \quad 1}$$

$$\boxed{0 \quad 1 \quad -1}$$

$$\boxed{1 \quad -1 \quad 1}$$

$$\boxed{-1 \quad 0 \quad 1} \qquad \boxed{1 \quad 0 \quad -1}$$

$$\boxed{-1 \quad 1 \quad -1}$$

$$\boxed{0 \quad -1 \quad 1}$$

$$\boxed{0 \quad 0 \quad -1}$$

We see that the twice anti-symmetric product of (1,0,0) contains (0,1,0), but the three times anti-symmetric product is (0,0,2). Thus we cannot build all the basic representations from (1,0,0). Nor can they all be built from (0,0,1). We must begin with both the (0,0,1) and (1,0,0) to generate all the basic representations.

Analogous considerations establish that a single representation will generate all representations for the $C_n$ algebras, but three initial representations are necessary for the $D_n$ algebras.

## Footnote

1. See JACOBSON, p 113.

## References

Representations are discussed at length in the appendix to DYNKIN III. For $SU(n)$, Young tableaux are the most effective procedure. They are explained in GEORGI. For a mathematical exposition, see BOERNER. For a very practical exposition, see SCHENSTED.

## Exercises

1. Find the weight scheme for the representations $(1, 0$ and $(0, 1)$ of $B_2$.

2. Find the weight scheme for $(1, 0)$ of $G_2$.

2. Find the weight scheme for $(1, 0)$ and $(0, 1)$ of $F_4$.

# XI. Casimir Operators and Freudenthal's Formula

One of the most familiar features of the analysis of SU(2) is the existence of an operator $T^2 = T_x^2 + T_y^2 + T_z^2$ which commutes with the generators, $T_x, T_y,$ and $T_z$. It is important to note that $T^2$ really has meaning only for representations, and not as an element of the Lie algebra since products like $t_x t_x$ are not defined for the algebra itself. Products like $T_x T_x$ are defined for representations since then $T_x$ is a linear transformation of a vector space into itself, and can be applied twice. We seek here the generalization of $T^2$ for an arbitrary simple Lie algebra.

It is well-known that $T^2 = \frac{1}{2}(T_+ T_- + T_- T_+) + T_z T_z$. This is the form which is easiest to relate to the forms we have used to describe Lie algebras in general. We might guess that the generalization will be roughly of the form

$$\mathcal{C} = \sum_{j,k} H_j M_{jk} H_k + \sum_{\alpha \neq 0} E_\alpha E_{-\alpha} \qquad (\text{XI.1})$$

where $H_j = H_{\alpha_j}$ and the $\alpha_j$ are a basis of simple roots. The matrix $M$ is to be determined by requiring that $\mathcal{C}$ commute with all the generators of the algebra. The normalizations of the $e_\alpha$ are chosen so $(e_\alpha, e_\beta) = \delta_{\alpha, -\beta}$, and thus

$$[e_\alpha, e_{-\alpha}] = h_\alpha \ , \tag{XI.2a}$$

$$[E_\alpha, E_{-\alpha}] = H_\alpha \ . \tag{XI.2b}$$

Let us define $N_{\alpha\beta}$ by

$$[e_\alpha, e_\beta] = N_{\alpha\beta} e_{\alpha+\beta} = -N_{\beta\alpha} e_{\alpha+\beta}. \tag{XI.3}$$

It is clear that $\mathcal{C}$ in Eq. (XI.1) already commutes with all the generators of the Cartan subalgebra since $[H_i, H_j] = 0$ and $[E_\alpha E_{-\alpha}, H_i] = 0$. It remains to calculate the commutator of $\mathcal{C}$ with $E_\beta$. We begin with the second term in $\mathcal{C}$:

$$\left[ \sum_{\alpha \neq 0} E_\alpha E_{-\alpha}, E_\beta \right] = \sum_{\substack{\alpha \neq 0 \\ \alpha \neq \beta}} E_\alpha N_{-\alpha,\beta} E_{\beta-\alpha}$$

$$+ \sum_{\substack{\alpha \neq 0 \\ \alpha \neq -\beta}} N_{\alpha\beta} E_{\alpha+\beta} E_{-\alpha} + E_\beta H_{-\beta} + H_{-\beta} E_\beta \ . \tag{XI.4}$$

We can obtain the necessary relation between the coefficients $N_{\alpha\beta}$ using the invariance of the Killing form:

$$(e_\alpha, [e_\beta, e_\gamma]) = N_{\beta,\gamma} \, \delta_{-\alpha,\beta+\gamma} = -N_{-\alpha-\beta,\beta} \, \delta_{\alpha+\beta,-\gamma}$$

$$= ([e_\alpha, e_\beta], e_\gamma)$$

$$= N_{\alpha,\beta} \delta_{\alpha+\beta,-\gamma}$$

$$N_{\alpha,\beta} = -N_{-\alpha-\beta,\beta} \ . \tag{XI.5}$$

Thus we have

$$\sum_{\substack{\alpha \neq 0 \\ \alpha \neq -\beta}} N_{\alpha\beta} E_{\alpha+\beta} E_{-\alpha} = \sum_{\substack{\alpha' \neq 0 \\ \alpha' \neq \beta}} N_{\alpha'-\beta,\beta} E_{\alpha'} E_{\beta-\alpha'}$$

$$= \sum_{\substack{\alpha' \neq 0 \\ \alpha' \neq \beta}} -N_{-\alpha',\beta} E_{\alpha'} E_{\beta-\alpha'} , \qquad \text{(XI.6)}$$

so the piece of the commutator of $E_\beta$ with $\mathcal{C}$ we have calculated is

$$\left[ \sum_{\alpha \neq 0} E_\alpha E_{-\alpha}, E_\beta \right] = E_\beta H_{-\beta} + H_{-\beta} E_\beta . \qquad \text{(XI.7)}$$

We now arrange the matrix $M$ so that the remainder of the commutator of $\mathcal{C}$ with $E_\beta$ just cancels this.

$$\left[ \sum_{j,k} H_j M_{jk} H_k, E_\beta \right] = \sum_{j,k} \left[ \langle \alpha_k, \beta \rangle H_j M_{jk} E_\beta + \langle \beta, \alpha_j \rangle M_{jk} E_\beta H_k \right] . \qquad \text{(XI.8)}$$

Now suppose that $\beta$ has the expansion in terms of simple roots

$$\beta = \sum_l k_l \alpha_l . \qquad \text{(XI.9)}$$

Then the coefficients are given by

$$k_l = \sum_j \langle \beta, \alpha_j \rangle \mathcal{A}_{jl}^{-1} \qquad \text{(XI.10)}$$

where the matrix $\mathcal{A}$ is

$$\mathcal{A}_{pq} = \langle \alpha_p, \alpha_q \rangle \qquad \text{(XI.11)}$$

and $\mathcal{A}^{-1}$ is its inverse. Now

$$H_\beta = \sum_l k_l H_l$$

$$= \sum_{j,k} \langle \beta, \alpha_j \rangle \mathcal{A}_{jk}^{-1} H_k \ . \tag{XI.12}$$

Thus if we take the matrix $M$ to be $\mathcal{A}^{-1}$, then the second portion of the commutator is

$$\left[ \sum_{j,k} H_j \mathcal{A}_{jk}^{-1} H_k, E_\beta \right] = H_\beta E_\beta + E_\beta H_\beta \ , \tag{XI.13}$$

just cancelling the first part. Altogether then

$$\mathcal{C} = \sum_{j,k} H_j \mathcal{A}_{jk}^{-1} H_k + \sum_{\alpha \neq 0} E_\alpha E_{-\alpha} \ . \tag{XI.14}$$

Consider SU(2) as an example. Our standard commutation relations are

$$[t_z, t_+] = t_+; \quad [t_z, t_-] = -t_-; \quad [t_+, t_-] = 2t_z \ , \tag{XI.15}$$

from which we find

$$(t_+, t_-) = \operatorname{Tr} \operatorname{ad} t_+ \operatorname{ad} t_- = 4 \ . \tag{XI.16}$$

Thus to obtain the normalization we have used in deriving the Casimir operator, we must set

$$t'_+ = \frac{1}{2} t_+ \ ; \quad t'_- = \frac{1}{2} t_- \ . \tag{XI.17}$$

so that

$$(t'_+, t'_-) = 1 \ . \tag{XI.18}$$

Now we regard $t'_+$ as the $e_\alpha$. The corresponding $h_\alpha$ is accordingly

$$h_\alpha = [t'_+, t'_-] = \frac{1}{2} t_z. \tag{XI.19}$$

It is straightforward to compute

$$\langle \alpha, \alpha \rangle = (h_\alpha, h_\alpha) = \frac{1}{4}(t_z, t_z) = \frac{1}{2}. \tag{XI.20}$$

It follows that the $1 \times 1$ matrix $M = \mathcal{A}^{-1}$ is simply 2. Altogether then, we find

$$\mathcal{C} = 2H_\alpha H_\alpha + E_\alpha E_{-\alpha} + E_{-\alpha} E_\alpha$$

$$= \frac{1}{2} T_z T_z + \frac{1}{4}(T_+ T_- + T_- T_+). \tag{XI.21}$$

This differs from the conventional SU(2) Casimir operator by an overall factor of $\frac{1}{2}$, a result simply of our need to establish some initial normalization in Eq. (XI.1). The importance of the Casimir operator is that since it commutes with all the generators, including the raising and lowering operators, it has the same value on every vector of an irreducible representation, since every such vector can be obtained by applying lowering operators to the highest weight vector. In fact, we can find the value of the Casimir operator on an irreducible representation by considering its action on the highest weight vector. Suppose the highest weight is $\Lambda$ and that $\phi_\Lambda$ is a vector with this weight. Then for every positive root $\alpha$ we know that $E_\alpha \phi_\Lambda = 0$ since otherwise it would have weight $\Lambda + \alpha$. On the other hand, we can compute $E_\alpha E_{-\alpha} \phi_\Lambda = (E_{-\alpha} E_\alpha + H_\alpha)\phi_\Lambda = \Lambda(h_\alpha)\phi_\Lambda = \langle \Lambda, \alpha \rangle \phi_\Lambda$, if $\alpha$ is positive. Thus we have

$$\mathcal{C}\phi_\Lambda = \sum_{j,k} H_j \mathcal{A}_{jk}^{-1} H_k \phi_\Lambda + \sum_{\alpha > 0} \langle \Lambda, \alpha \rangle \phi_\Lambda$$

$$= \sum_{j,k} \langle \Lambda \alpha_j \rangle \mathcal{A}_{jk}^{-1} \langle \Lambda, \alpha_k \rangle \phi_\Lambda + \sum_{\alpha > 0} \langle \Lambda, \alpha \rangle \phi_\Lambda. \tag{XI.22}$$

Thus on this irreducible representation

$$\mathcal{C} = \sum_{j,k} \langle \Lambda \alpha_j \rangle \mathcal{A}_{jk}^{-1} \langle \Lambda, \alpha_k \rangle + \sum_{\alpha > 0} \langle \Lambda, \alpha \rangle$$

$$= \langle \Lambda, \Lambda \rangle + \langle \Lambda, 2\delta \rangle \tag{XI.23}$$

where we have introduced the element of $H_0^*$

$$\delta = \frac{1}{2} \sum_{\alpha > 0} \alpha . \tag{XI.24}$$

A few comments are in order concerning normalizations. We have derived our scalar product from the Killing form. As we saw in Chapter IV, all invariant bilinear forms are proportional to the Killing form if the algebra is simple. Suppose we define a second form by

$$(x, y)' = c(x, y) = c\,\mathrm{Tr}\,\mathrm{ad}\,x\,\mathrm{ad}\,y . \tag{XI.22}$$

Now since $h_\rho$ is defined by

$$(h_\rho, k) = \rho(k) \tag{XI.23}$$

we define $h_\rho'$ by

$$(h_\rho', k)' = \rho(k) \tag{XI.24}$$

so that

$$h_\rho' = \frac{1}{c} h_\rho \tag{XI.25}$$

and

$$\langle \rho, \tau \rangle' \equiv (h'_\rho, h'_\tau)' = \frac{1}{c} \langle \rho, \tau \rangle . \tag{XI.26}$$

We now compare the commutation relations as expressed using the two different scalar products. We have

$$[h_\alpha, e_\beta] = \beta(h_\alpha)e_\beta = (h_\beta, h_\alpha)e_\beta = \langle \alpha, \beta \rangle e_\beta \tag{XI.27}$$

which becomes

$$[h'_\alpha, e_\beta] = \beta(h'_\alpha)e_\beta = (h'_\alpha, h'_\beta)'e_\beta = \langle \alpha, \beta \rangle' e_\beta . \tag{XI.28}$$

Thus the commutation relations look the same for this new scalar product. A new Casimir operator (which is just a multiple of the old one) can be chosen so that its value is just $\langle \Lambda, \Lambda + 2\delta \rangle'$. In this way, we can choose a scalar product with any desired normalization and have the computations go through just as before. For some applications, it is traditional to use a scalar product which gives the largest root a length squared equal to 2. We indicate this scalar product by $\langle \ , \ \rangle_2$.

One particular way to choose an invariant bilinear form is to take the trace of two representation matrices. That is, if $\phi$ is a representation $\phi(e_\alpha) = E_\alpha$, etc., then we define

$$((x, y)) = \text{Tr} \, \phi(x)\phi(y) . \tag{XI.29}$$

The invariance of this form follows from the invariance of traces under cyclic permutation. We know then that $(( \ , \ ))$ is proportional to the Killing form and to the form $( \ , \ )_2$ which yields $\langle \ , \ \rangle_2$. The constant of proportionality to the latter is called the **index of the representation**, $l_\phi$:

$$((x, y)) = l_\phi(x, y)_2 . \tag{XI.30}$$

Now we can evaluate $l_\phi$ by considering $\mathcal{C}$ with the normalization appropriate to $( \ , \ )_2$. For a representation with highest weight $\Lambda$, $\mathcal{C} = \langle \Lambda, \Lambda + 2\delta \rangle_2$.

If we take $\text{Tr}\,\mathcal{C}$, we get $N_\Lambda \langle \Lambda, \Lambda + 2\delta \rangle_2$, where $N_\Lambda$ is the dimension of the representation. On the other hand, replacing $(( \ , \ ))$ by $l_\phi( \ , \ )_2$ yields $l_\phi N_{adj}$ where $N_{adj}$ is the dimension of the algebra, that is, the dimension of the adjoint representation. Thus

$$l_\phi = \frac{N_\Lambda \langle \Lambda, \Lambda + 2\delta \rangle_2}{N_{adj}} \ . \tag{XI.31}$$

We shall see some applications of the index in later chapters.

One particular application of the Casimir operator is in deriving **Freudenthal's Recursion Formula** for the dimensionality of a weight space. Previously, we developed an algorithm for determining all the weights of an irreducible representation, but without ascertaining the dimensionality of each weight space, an omission which we now rectify. Subsequently, this result will be used to derive Weyl's formula for the dimension of an irreducible representation.

We consider an irreducible representation whose highest weight is $\Lambda$ and seek the dimensionality of the space with weight $M$. Now we know the constant value of $\mathcal{C}$ on the whole carrier space of the representation, so we can calculate the trace of $\mathcal{C}$ restricted to the space with weight $M$:

$$\text{Tr}_M \mathcal{C} = n_M \langle \Lambda, \Lambda + 2\delta \rangle \ . \tag{XI.32}$$

Here $n_M$ is the dimensionality of the space with weight $M$, that is, the quantity we wish to compute. Now we calculate the same quantity another way. The first part of $\mathcal{C}$ gives us

$$\text{Tr}_M \sum_{j,k} H_j \mathcal{A}_{jk}^{-1} H_k = \sum_{j,k} \langle \alpha_j, M \rangle \mathcal{A}_{jk}^{-1} \langle \alpha_k, M \rangle n_M$$

$$= n_M \langle M, M \rangle \ . \tag{XI.33}$$

What remains is

$$\operatorname{Tr}_M \sum_{\alpha>0}(E_\alpha E_{-\alpha} + E_{-\alpha}E_\alpha) , \tag{XI.34}$$

where our normalization is $(e_\alpha, e_{-\alpha}) = 1$ so

$$[E_\alpha, E_{-\alpha}] = H_\alpha, \qquad [H_\alpha, E_\alpha] = \langle \alpha, \alpha \rangle E_\alpha . \tag{XI.35}$$

Now usually for SU(2) we have

$$[T_+, T_-] = 2T_z, \qquad [T_z, T_+] = T_+ \tag{XI.36}$$

and

$$T^2 = T_z^2 + \frac{1}{2}[T_+T_- + T_-T_+] . \tag{XI.37}$$

We want to exploit our understanding of SU(2) so we consider the SU(2) generated by $E_\alpha, E_{-\alpha}$, and $H_\alpha$. The correspondence which gives the right normalization is

$$T_z = \frac{H_\alpha}{\langle \alpha, \alpha \rangle}, \qquad T_+ = \sqrt{\frac{2}{\langle \alpha, \alpha \rangle}}E_\alpha , \qquad T_- = \sqrt{\frac{2}{\langle \alpha, \alpha \rangle}}E_{-\alpha} . \tag{XI.38}$$

Now consider the weight space associated with the weight $M$. The full irreducible representation contains, in general, many irreducible representations of the SU(2) associated with the root $\alpha$. We can pick a basis for the weight space for weight $M$ so that each basis vector belongs to a distinct irreducible representation of the SU(2). Each such irreducible representation is characterized by an integer or half-integer, $t$ which is the maximal eigenvalue of $T_z$. Moreover, the usual Casimir operator, Eq. (XI.37), then has the value $t(t+1)$. If $\phi_t$ is an appropriate weight vector then we can write

$$\left[ \frac{H_\alpha H_\alpha}{\langle \alpha, \alpha \rangle^2} + \frac{1}{\langle \alpha, \alpha \rangle} E_\alpha E_{-\alpha} + \frac{1}{\langle \alpha, \alpha \rangle} E_{-\alpha} E_\alpha \right] \phi_t = t(t+1)\phi_t \tag{XI.39}$$

so that

$$\left[ E_\alpha E_{-\alpha} + E_{-\alpha} E_\alpha \right] \phi_t = \langle \alpha, \alpha \rangle t(t+1)\phi_t - \frac{\langle M, \alpha \rangle^2}{\langle \alpha, \alpha \rangle} \phi_t \tag{XI.40}$$

where we have used the fact that $\phi_t$ has weight $M$. The particular weight vector $\phi_t$ belongs to a series of weight vectors which form a basis for an irreducible representation of the $SU(2)$ described above. Suppose the highest weight in this series is $M + k\alpha$, and the associated weight vector is $\phi_{M+k\alpha}$. Then

$$T_z \phi_{M+k\alpha} = t\phi_{M+k\alpha} = \frac{H_\alpha}{\langle \alpha, \alpha \rangle} \phi_{M+k\alpha} = \frac{\langle \alpha, M + k\alpha \rangle}{\langle \alpha, \alpha \rangle} \phi_{M+k\alpha} . \tag{XI.41}$$

Thus we can indentify

$$t = \frac{\langle \alpha, M + k\alpha \rangle}{\langle \alpha, \alpha \rangle} . \tag{XI.42}$$

This result can now be inserted in Eq. (XI.39) to find

$$\left[ E_\alpha E_{-\alpha} + E_{-\alpha} E_\alpha \right] \phi_t = [k(k+1)\langle \alpha, \alpha \rangle + (2k+1)\langle M, \alpha \rangle]\phi_t . \tag{XI.43}$$

Each of our basis vectors for the space with weight $M$ has associated with it a value of $k$. In fact, more than one basis vector may have the same value of $k$. A moment's reflection reveals that the number of such basis vectors is precisely the difference between the dimension of the space with weight $M + (k+1)\alpha$ and that with weight $M + k\alpha$. Accordingly, we write

$$\text{Tr}_M \sum_{\alpha > 0} \left[ E_\alpha E_{-\alpha} + E_{-\alpha} E_\alpha \right]$$

$$= \sum_{k \geq 0} \left( n_{M+k\alpha} - n_{M+(k+1)\alpha} \right) [k(k+1)\langle \alpha, \alpha \rangle + (2k+1)\langle M, \alpha \rangle]$$

$$= n_M \langle M, \alpha \rangle + \sum_{k=1}^{\infty} n_{M+k\alpha}[2k\langle \alpha, \alpha \rangle + 2\langle M, \alpha \rangle] . \tag{XI.44}$$

Combining this result with Eqs. (XI.32) and (XI.33), we find

$$\mathrm{Tr}_M \mathcal{C} = n_M \langle \Lambda, \Lambda + 2\delta \rangle$$

$$= n_M \langle M, M \rangle$$

$$+ \sum_{\alpha > 0} \left[ n_M \langle M, \alpha \rangle + \sum_{k=1}^{\infty} 2 n_{M+k\alpha} \langle M + k\alpha, \alpha \rangle \right] . \qquad \text{(XI.45)}$$

This relation may be solved for $n_M$ in terms of the higher weights:

$$n_M = \frac{\sum_{\alpha > 0} \sum_{k=1}^{\infty} 2 n_{M+k\alpha} \langle M + k\alpha, \alpha \rangle}{\langle \Lambda + M + 2\delta, \Lambda - M \rangle} . \qquad \text{(XI.46)}$$

The highest weight always has a space of dimension one. Using Freudenthal's formula, Eq. (XI.45), we can determine the dimensionality of the spaces of the lower weights successively. The denominator is most easily evaluated by expressing the first factor by its Dynkin coefficients and the second factor in terms of simple roots. As we shall demonstrate later, the Dynkin coefficients of $\delta$ are $(1, 1, \ldots)$. Since $\Lambda$ and $M$ appear in the table of weights expressed in Dynkin coefficients, it is easy then to find $\Lambda + M + 2\delta$. Similarly, $\Lambda - M$ is easily determined from the table of weights. If the Dynkin coefficients of $\Lambda + M + 2\delta$ are

$$(\Lambda + M + 2\delta) = (a_1, a_2, \ldots) \qquad \text{(XI.47)}$$

and

$$\Lambda - M = \sum_i k_i \alpha_i \qquad \text{(XI.48)}$$

then

$$\langle \Lambda + M + 2\delta, \Lambda - M \rangle = \sum_i a_i k_i \frac{1}{2} \langle \alpha_i, \alpha_i \rangle . \qquad \text{(XI.49)}$$

The quantity in the numerator can be evaluated fairly easily as well. For a given positive root, $\alpha$, we check to see whether $M' = M + k\alpha$ is also a weight. If it is, then $M'$ and $M$ lie in a string of weights separated by $\alpha$'s. Let the highest and lowest weights in the string be $M' + p\alpha$ and $M' - m\alpha$. Then by Eq. (V.22),

$$2\langle M', \alpha \rangle = (m - p)\langle \alpha, \alpha \rangle . \tag{XI.50}$$

Let us consider an application of the Freudenthal formula to SU(3). The 27-dimensional representation has the Dynkin representation (2,2). It is the representation with highest weight in the product of two adjoint representations.

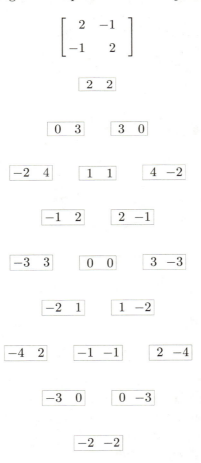

$$\begin{bmatrix} 2 & -1 \\ -1 & 2 \end{bmatrix}$$

First note that the weights (0,3) and (3,0) clearly are one dimensional since there is only one way to reach them by lowering operators from the highest weight. Thus the first ambiguity is for the weight (1,1). We compute with M=(1,1)

$$\Lambda + M + 2\delta = (5,5)$$

$$\Lambda - M = \alpha_1 + \alpha_2$$

$$\langle \Lambda + M + 2\delta, \Lambda - M \rangle = 10 \cdot \frac{1}{2} \langle \alpha_1, \alpha_1 \rangle \qquad \text{(XI.51)}$$

where we have used the relation $\langle \alpha_1, \alpha_1 \rangle = \langle \alpha_2, \alpha_2 \rangle$.

To compute the numerator, we remember that there are three positive roots $\alpha_1, \alpha_2$, and $\alpha_1 + \alpha_2$. The weight preceding (1,1) by $\alpha_1$ is (3,0). For this weight, m=3 and p=0. Similarly for (0,3) which precedes (1,1) by $\alpha_2$. The weight preceding (1,1) by $\alpha_1 + \alpha_2$ is (2,2). For this weight, m=4 and p =0. Remembering that all the roots of SU(3) have the same size, we have for the numerator $(3+3+4)\langle \alpha_1, \alpha_1 \rangle$ and thus

$$N_{(1,1)} = 2 \ . \qquad \text{(XI.52)}$$

For the weight (2,-1) we have

$$\Lambda + M + 2\delta = (6,3)$$

$$\Lambda - M = \alpha_1 + 2\alpha_2$$

$$\langle \Lambda + M + 2\delta, \Lambda - M \rangle = (6+6) \cdot \frac{1}{2} \cdot \langle \alpha_1, \alpha_1 \rangle \ . \qquad \text{(XI.53)}$$

The numerator receives a contribution of $4\langle \alpha_1, \alpha_1 \rangle$ from the root $\alpha_1$. For the root $\alpha_2$ there are two preceding weights to consider. The weight (0,3) contributes $3\langle \alpha_1, \alpha_1 \rangle$. The weight (1,1) contributes $2(2-1)\langle \alpha_1, \alpha_1 \rangle$, where the factor of two comes from the dimensionality of the weight space for (1,1). For the root $\alpha_1 + \alpha_2$, there is a contribution $3\langle \alpha_1, \alpha_1 \rangle$. Altogether, then,

$$N_{(2,-1)} = \frac{4 + 3 + 2 + 3}{12 \cdot \frac{1}{2}} = 2 \ . \qquad \text{(XI.54)}$$

# References

The derivation of Freudenthal's formula follows that given by JACOBSON, pp. 243–249. The index of a representation is developed by DYNKIN II, see in particular pp. 130–134.

# Exercises

1. Find the index of the seven dimensional representation of $G_2$.

2. Find the dimensionalities of all the weight spaces of the 27-dimensional representation of SU(3).

3. Show that the index of the $k$ dimensional representation of $SU(2)$ is $(k - 1)k(k + 1)/6$.

# XII. The Weyl Group

The irreducible representations of SU(2) manifest a very obvious symmetry: for every state with $T_z = m$ there is a state with $T_z = -m$. This symmetry is the source of a more complex symmetry in larger algebras. The SU(2) representations are symmetric with respect to reflection about their centers. The larger algebras have reflection symmetries and the group generated by these reflections is called the **Weyl group**.

Consider an irreducible representation of a simple Lie algebra. Now if $\alpha$ is a root of the algebra, we can consider the SU(2) generated by $e_\alpha, e_{-\alpha}$, and $h_\alpha$. The representation of the full algebra will in general be a reducible representation of this SU(2). Let $M$ be some particular weight and consider the weights and weight spaces associated with $\ldots, M + \alpha, M, M - \alpha, \ldots$. These together form some reducible representation of the SU(2). Under the SU(2) reflection, this representation is mapped into itself. Moreover, since each weight space has a basis in which each element belongs to a distinct SU(2) representation, it is clear that the reflection will map one weight space into another of the same dimension. What is the relation

98

between the original weight and the one into which it is mapped? This is easily inferred from geometry. The portion of $M$ parallel to $\alpha$ is $\alpha\langle M, \alpha\rangle/\langle \alpha, \alpha\rangle$ and the portion perpendicular to it is then $M - \alpha\langle M, \alpha\rangle/\langle \alpha, \alpha\rangle$. The reflection changes the sign of the former and leaves unchanged the latter. Thus the reflection of the weight can be written

$$S_\alpha(M) = M - 2\frac{\langle M, \alpha\rangle}{\langle \alpha, \alpha\rangle}\alpha \, , \qquad\qquad \text{(XII.1)}$$

where $S_\alpha$ is an operator acting on the space of weights, $H_0^*$. It maps weights into weights whose weight spaces are of the same dimension. If we let $\alpha$ range over all the roots of the algebra we get a collection of reflections. By taking all combinations of these reflections applied successively, we obtain the Weyl group.

The 27-dimensional representation of SU(3) provides a good example of the symmetry at hand. The $Y = 0$ subspace contains three SU(2) multiplets, one with $T = 2$, one with $T = 1$, and one with $T = 0$. The $Y = 1$ subspace contains two SU(2) multiplets, one with $T = 3/2$ and one with $T = 1/2$. The $Y = 2$ subspace has $T = 1$. The SU(2) reflection maps the weight diagram into itself, preserving the dimensionality of each weight space.

Rather than consider all the $S_\alpha$, it turns out that it suffices to consider just those $S_\alpha$ where $\alpha \in \Pi$. These will also generate the full Weyl group. For SU(3) we find

$$S_{\alpha_1} : \quad \alpha_1 \to -\alpha_1$$

$$\alpha_2 \to \alpha_1 + \alpha_2 = \alpha_3$$

$$S_{\alpha_2} : \quad \alpha_1 \to \alpha_1 + \alpha_2 = \alpha_3$$

$$\alpha_2 \to -\alpha_2 \, . \qquad\qquad \text{(XII.2)}$$

The full Weyl group for SU(3) has six elements.

We shall not need to know much about the Weyl group for specific algebras. The utility of the Weyl group is that it enables us to prove quite general propositions without actually having to consider the details of representations since it permits the exploitation of their symmetries.

Let us prove a number of useful facts about the Weyl group. First, the Weyl group is a set of orthogonal transformations on the weight space. Orthogonal transformations are those which preserve the scalar product. It is intuitively clear that reflections have this property. To prove this for the full Weyl group it suffices to prove it for the $S_\alpha$ which generate it. We have

$$\langle S_\alpha M, S_\alpha N \rangle = \langle M - 2\frac{\langle M, \alpha \rangle}{\langle \alpha, \alpha \rangle}\alpha, N - 2\frac{\langle N, \alpha \rangle}{\langle \alpha, \alpha \rangle}\alpha \rangle$$

$$= \langle M, N \rangle . \tag{XII.3}$$

We know that the Weyl group maps weights into weights, so by taking the adjoint representation, we see that it maps roots into roots. The particular reflections $S_\alpha$ where $\alpha$ is simple have a special property. Certainly $S_\alpha(\alpha) = -\alpha$. For every other positive root, $\beta \in \Sigma^+$, $S_\alpha(\beta)$ is positive. To see this, express

$$\beta = \sum_j k_j \alpha_j . \tag{XII.4}$$

Now $\alpha$ is one of the $\alpha_j$'s, say $\alpha = \alpha_1$. Thus

$$S_{\alpha_1}(\beta) = \sum_j k_j \alpha_j - 2\alpha_1 \sum_j k_j \frac{\langle \alpha_j, \alpha_1 \rangle}{\langle \alpha_1, \alpha_1 \rangle}$$

$$= \sum_{j>1} k_j \alpha_j + \alpha_1 \times \text{something} . \tag{XII.5}$$

Since $\beta \neq \alpha = \alpha_1$, some $k_j \neq 0$ for $j > 1$. Thus the root $S_\alpha(\beta)$ has some positive coefficient in its expansion in terms of simple roots. But this is enough to establish that all the coefficients are positive and hence so is the root.

The Weyl group provides the means to prove the relation used in the preceding section, that the Dynkin coefficients of $\delta = \frac{1}{2}\sum_{\alpha>0} \alpha$ are all unity. Let $\alpha_i$ be one of the simple roots. By the orthogonality of the Weyl reflections,

$\langle S_{\alpha_i}\delta, \alpha_i \rangle = \langle \delta, -\alpha_i \rangle$. On the other hand, $S_{\alpha_i}$ interchanges all the positive roots except $\alpha_i$ itself, so $S_{\alpha_i}\delta = \delta - \alpha_i$. Thus

$$\langle \delta - \alpha_i, \alpha_i \rangle = \langle \delta, -\alpha_i \rangle$$

$$2\langle \delta, \alpha_i \rangle = \langle \alpha_i, \alpha_i \rangle \tag{XII.6}$$

as we wished to show.

Finally, consider all the weights $M'$ which can be obtained by acting on the weight $M$ with an element $S \in W$, the Weyl group. We claim that the $M'$ which is the highest has Dynkin coefficients which are all non-negative. Suppose $M^*$ is the highest of these weights $SM$, and further suppose that the Dynkin coefficient $2\langle M^*, \alpha_i \rangle / \langle \alpha_i, \alpha_i \rangle < 0$. Then $S_{\alpha_i}M^* = M^* - 2\alpha_i \langle M^*, \alpha_i \rangle / \langle \alpha_i, \alpha_i \rangle$ is an even higher weight, providing a contradiction.

# References

The Weyl group is covered by JACOBSON, pp. 240–243.

# Exercise

1. Find the elements of the Weyl group for $G_2$ and their multiplication table.

# XIII. Weyl's Dimension Formula

In this Section we derive the celebrated formula of Weyl for the dimensionality of an irreducible representation of a simple Lie algebra in terms of its highest weight. Our derivation is essentially that of Jacobson, which is based on the technique of Freudenthal.

We shall be considering functions defined on $H_0^*$. Instead of parameterizing elements of $H_0^*$ in terms of the simple roots, it is convenient to over-parameterize by writing $\rho \in H_0^*$ as

$$\rho = \sum_{\alpha \in \Sigma} \rho_\alpha \alpha \ . \tag{XIII.1}$$

We can define the action of an element, $S$, of the Weyl group on a function of $\rho$ by

$$(SF)(\rho) = F(S^{-1}\rho) \ . \tag{XIII.2}$$

As an example, consider the function known as the **character of the represen-tation**

$$\chi(\rho) = \sum_M n_M \exp\langle M, \rho \rangle \qquad (XIII.3)$$

where the sum is over all the weights of a representation and $n_M$ is the dimensionality of the weight space for $M$. Now we calculate the action of $S \in W$ on $\chi$:

$$(S\chi)(\rho) = \sum_M n_M \exp\langle M, S^{-1}\rho \rangle$$

$$= \sum_M n_M \exp\langle SM, \rho \rangle$$

$$= \sum_M n_M \exp\langle M, \rho \rangle \ . \qquad (XIII.4)$$

Here we have used the orthogonality property of the Weyl group and the relation $n_M = n_{SM}$. Thus we see that $S\chi = \chi$, that is $\chi$ is invariant under the Weyl group.

Consider next the function

$$Q(\rho) = \prod_{\alpha > 0} [\exp(\tfrac{1}{2}\langle \alpha, \rho \rangle) - \exp(-\tfrac{1}{2}\langle \alpha, \rho \rangle)] \ . \qquad (XIII.5)$$

We want to determine the behavior of this function when acted upon by elements of the Weyl group. It suffices to determine the effect of the $S_i = S_{\alpha_i}$, the reflections associated with simple roots.

$$(S_iQ)(\rho) = \prod_{\alpha > 0} \left[ \exp\left(\tfrac{1}{2}\langle \alpha, S_i^{-1}\rho \rangle\right) - \exp\left(-\tfrac{1}{2}\langle \alpha, S_i^{-1}\rho \rangle\right) \right]$$

$$= \prod_{\alpha > 0} \left[ \exp\left(\tfrac{1}{2}\langle S_i\alpha, \rho \rangle\right) - \exp\left(-\tfrac{1}{2}\langle S_i\alpha, \rho \rangle\right) \right] \ . \qquad (XIII.6)$$

We have already seen that $S_i$ interchanges all the positive roots except $\alpha_i$ whose sign it changes. Thus we see directly that

$$(S_iQ)(\rho) = -Q(\rho) . \tag{XIII.7}$$

Now $S_i$ reverses the sign of $\alpha_i$, but leaves every vector orthogonal to $\alpha_i$ unchanged. Thus $\det S_i = -1$ and we can write

$$(S_iQ) = (\det S_i)Q . \tag{XIII.8}$$

Indeed, every $S \in W$ is a product of $S_i$'s, so

$$SQ = \det SQ . \tag{XIII.9}$$

Functions with this property are called **alternating**.

We can make alternating functions by applying the operator

$$\sigma = \sum_{S \in W} (\det S)S \tag{XIII.10}$$

for we have

$$S'\sigma = \sum_{S \in W} S'(\det S)S$$

$$= \sum_{S \in W} \det S' \det(S'S)S'S$$

$$= \det S'\sigma . \tag{XIII.11}$$

It is convenient to find a representation of the alternating function $Q(\rho)$ in the form $\sigma F(\rho)$. From the definition of $Q(\rho)$ it is clear that there must be an expansion of the form

$$Q(\rho) = \sigma \sum_\beta c_\beta \exp\langle \delta - \beta, \rho \rangle \qquad \text{(XIII.12)}$$

where $\delta$ is half the sum of all the positive roots and where $\beta$ is a sum of distinct positive roots. Now in such an expansion it is redundant to include both $\sigma \exp\langle M, \rho \rangle$ and $\sigma \exp\langle SM, \rho \rangle$ since they are equal up to a factor $\det S$. We have already seen that among all the $SM$, $S \in W$, the largest one has only non-negative Dynkin coefficients. Thus we need only consider terms where $\delta - \beta$ has only non-negative Dynkin coefficients. In fact, we can restrict this further because if $M$ has a Dynkin coefficient which is zero, then $\sigma \exp\langle M, \rho \rangle = 0$. This is easy to establish since if $M(h_i) = 0$, then $S_i M = M$ and $S_i \sigma \exp\langle M, \rho \rangle = \sigma \exp\langle S_i M, \rho \rangle = \sigma \exp\langle M, \rho \rangle = \det S_i \sigma \exp\langle M, \rho \rangle = -\sigma \exp\langle M, \rho \rangle = 0$. However, we have seen that $\delta$ has Dynkin coefficients which are all unity. Now since $\beta$ is a sum of positive roots it cannot have only negative Dynkin coefficients. Thus we see that the sum, Eq. (XIII.12) need include only $\beta = 0$. Comparing the coefficients of $\exp\langle \delta, \rho \rangle$ we see that $c_{\beta=0} = 1$ so

$$Q(\rho) = \prod_{\alpha>0} [\exp \frac{1}{2}\langle \alpha, \rho \rangle - \exp -\frac{1}{2}\langle \alpha, \rho \rangle]$$

$$= \sum_{S \in W} (\det S) \exp\langle S\delta, \rho \rangle . \qquad \text{(XIII.13)}$$

We shall now use these results to analyze the character. We begin with the Freudenthal recursion relation, Eq. (XI.45), together with $\sum_{k=-\infty}^{\infty} n_{M+k\alpha}\langle M + k\alpha, \alpha \rangle = 0$ :

$$[\langle \Lambda + \delta, \Lambda + \delta \rangle - \langle \delta, \delta \rangle - \langle M, M \rangle] n_M$$

$$= \sum_{\alpha \neq 0} \sum_{k=0}^{\infty} n_{M+k\alpha}\langle M + \kappa\alpha, \alpha \rangle. \qquad \text{(XIII.14)}$$

Mulitplying by $\exp\langle M, \rho \rangle$ and summing over $M$, we have

$$[\langle \Lambda + \delta, \Lambda + \delta \rangle - \langle \delta, \delta \rangle]\chi - \sum_M n_m \langle M, M \rangle \exp\langle M, \rho \rangle$$

$$= \sum_M \sum_{\alpha \neq 0} \sum_{k=0}^{\infty} n_{M+\kappa\alpha} \langle M + \kappa\alpha, \alpha \rangle \exp\langle M, \rho \rangle . \quad \text{(XIII.15)}$$

Remembering that (see Eq. (IV.3))

$$\langle M, N \rangle = \sum_{\alpha \in \Sigma} \alpha(h_M)\alpha(h_N) = \sum_{\alpha \in \Sigma} \langle M, \alpha \rangle \langle \alpha, N \rangle , \quad \text{(XIII.16)}$$

we derive the relations

$$\frac{\partial}{\partial \rho_\alpha} \exp\langle M, \rho \rangle = \langle \alpha, M \rangle \exp\langle M, \rho \rangle , \quad \text{(XIII.17}a\text{)}$$

$$\sum_{\alpha \in \Sigma} \frac{\partial^2}{\partial \rho_\alpha^2} \exp\langle M, \rho \rangle = \langle M, M \rangle \exp\langle M, \rho \rangle . \quad \text{(XIII.17}b\text{)}$$

Inserting this into Eq. (XIII.15) gives

$$[\langle \Lambda + \delta, \Lambda + \delta \rangle - \langle \delta, \delta \rangle - \sum_{\alpha \in \Sigma} \frac{\partial^2}{\partial \rho_\alpha^2}]\chi$$

$$= \sum_M \sum_{\alpha \neq 0} \sum_{k=0}^{\infty} n_{M+\kappa\alpha} \langle M + \kappa\alpha, \alpha \rangle \exp\langle M, \rho \rangle . \quad \text{(XIII.18)}$$

To analyze the right hand side of Eq. (XIII.18), let us first fix $\alpha \neq 0$ and consider the SU(2) generated by $E_\alpha, E_{-\alpha}$, and $H_\alpha$. The irreducible representation of the full algebra is, in general, a reducible representation of this SU(2). The dimensionality $n_M$ is just the number of SU(2)-irreducible representations present at the weight $M$. Thus we can proceed by calculating the contribution of each SU(2)-irreducible representation to the sum for fixed $M$ and $\alpha$. The string of weights containing $M$ which correspond to an SU(2) irreducible representation are distributed symmetrically about a center point, $M_0$ which can be expressed in terms of the highest weight in the sequence, $M^*$, as

$$M^0 = M^* - \alpha \frac{\langle M^*, \alpha \rangle}{\langle \alpha, \alpha \rangle} . \tag{XIII.19}$$

Note that $\langle M^0, \alpha \rangle = 0$.

Thus each weight in the sequence is of the form $M^0 + m\alpha$ where $m$ is an integer or half-integer. The range of $m$ is from $-j$ to $j$, where again $j$ is an integer or half-integer. Now we can write the contribution of a single SU(2) irreducible representation to the sum as

$$\sum_M \sum_{k=0}^{\infty} \langle M + \kappa\alpha, \alpha \rangle \exp\langle M, \rho \rangle$$

$$= \sum_m \sum_{k=0}^{j-m} \langle M^0 + m\alpha + k\alpha, \alpha \rangle \exp\langle M^0 + m\alpha, \rho \rangle$$

$$= \sum_m \sum_{k=0}^{j-m} \langle \alpha, \alpha \rangle (m + k) \exp\langle M^0, \rho \rangle \exp(m\langle \alpha, \rho \rangle)$$

$$= \langle \alpha, \alpha \rangle \exp\langle M^0, \rho \rangle \sum_m \sum_{k=0}^{j-m} (m + k) \exp(m\langle \alpha, \rho \rangle) .$$

$$\tag{XIII.20}$$

The double sum is most easily evaluated by multiplying first by $\exp\langle\alpha,\rho\rangle - 1$:

$$\sum_{m=-j}^{j} \sum_{k=0}^{j-m} (k+m) \exp\langle m\alpha,\rho\rangle (\exp\langle\alpha,\rho\rangle - 1)$$

$$= \sum_{m=-j}^{j} \sum_{k=0}^{j-m} (m+k) \exp\langle(m+1)\alpha,\rho\rangle$$

$$- \sum_{m=-j-1}^{j-1} \sum_{k=0}^{j-m-1} (m+k+1) \exp\langle(m+1)\alpha,\rho\rangle$$

$$= j \exp\langle(j+1)\alpha,\rho\rangle + \sum_{m=-j}^{j-1} [j-(j-m)] \exp\langle(m+1)\alpha,\rho\rangle$$

$$+ \sum_{k=0}^{2j} (k-j) \exp\langle -j\alpha,\rho\rangle$$

$$= \sum_{m=-j}^{j} m \exp\langle(m+1)\alpha,\rho\rangle \ . \tag{XIII.21}$$

Thus the contribution of one $SU(2)$ irreducible representation to the original sum is

$$\langle\alpha,\alpha\rangle \exp\langle M^0,\rho\rangle \sum_{m=-j}^{j} m \exp\langle(m+1)\alpha,\rho\rangle [\exp\langle\alpha,\rho\rangle - 1]^{-1}$$

$$= \sum_{m=-j}^{j} \langle M,\alpha\rangle \exp\langle M+\alpha,\rho\rangle [\exp\langle\alpha,\rho\rangle - 1]^{-1} \ . \tag{XIII.22}$$

Summing over all $SU(2)$ irreducible representations, and over all the roots, we have

$$[\langle\Lambda+\delta,\Lambda+\delta\rangle - \langle\delta,\delta\rangle - \sum_{\alpha\in\Sigma} \frac{\partial^2}{\partial\rho_\alpha^2}]\chi$$

$$= \sum_{\alpha\neq 0} \sum_{M} n_M \langle\alpha,M\rangle \exp\langle M+\alpha,\rho\rangle [\exp\langle\alpha,\rho\rangle - 1]^{-1} \ . \tag{XIII.23}$$

From the definition of $Q$, we see that

$$\prod_{\alpha \neq 0} [\exp\langle \alpha, \rho \rangle - 1] = \eta Q^2(\rho) \qquad \text{(XIII.24)}$$

where $\eta$ is $+1$ if the number of positive roots is even and $-1$ if it is odd. Thus

$$\frac{\partial}{\partial \rho_\beta} \log \eta Q^2(\rho) = \sum_{\alpha \neq 0} \frac{\exp\langle \alpha, \rho \rangle}{\exp\langle \alpha, \rho \rangle - 1} \langle \alpha, \beta \rangle \qquad \text{(XIII.25)}$$

and

$$\frac{\partial}{\partial \rho_\beta} \exp\langle M, \rho \rangle = \langle \beta, M \rangle \exp\langle M, \rho \rangle \qquad \text{(XIII.26)}$$

so

$$\sum_M \sum_{\alpha \neq 0} n_M \langle \alpha, M \rangle \exp\langle M + \alpha, \rho \rangle [\exp\langle \alpha, \rho \rangle - 1]^{-1}$$

$$= \sum_\beta \frac{\partial}{\partial \rho_\beta} \log \eta Q^2(\rho) \frac{\partial}{\partial \rho_\beta} \chi$$

$$= 2Q^{-1} \sum_\beta \frac{\partial}{\partial \rho_\beta} Q \frac{\partial}{\partial \rho_\beta} \chi$$

$$= Q^{-1} \left[ \sum_\beta \frac{\partial^2}{\partial \rho_\beta^2} (Q\chi) - Q \sum_\beta \frac{\partial^2}{\partial \rho_\beta^2} \chi - \chi \sum_\beta \frac{\partial^2}{\partial \rho_\beta^2} Q \right] . \qquad \text{(XIII.27)}$$

Combining these results, we have the differential equation

$$\langle \Lambda + \delta, \Lambda + \delta \rangle Q\chi = \chi \left[ \langle \delta, \delta \rangle - \sum_\beta \frac{\partial^2}{\partial \rho_\beta^2} \right] Q + \sum_\beta \frac{\partial^2}{\partial \rho_\beta^2} Q\chi . \qquad \text{(XIII.28)}$$

From the relation

$$Q(\rho) = \sum_{S \in W} (\det S) \exp\langle S\delta, \rho\rangle , \qquad (\text{XIII.29})$$

it follows that

$$\sum_{\beta} \frac{\partial^2}{\partial \rho_\beta^2} Q(\rho) = \langle \delta, \delta\rangle Q(\rho) , \qquad (\text{XIII.30})$$

where we have used the orthogonality of the $S$'s. Altogether then we have

$$\sum_{\beta} \frac{\partial^2}{\partial \rho_\beta^2} Q\chi = \langle \Lambda + \delta, \Lambda + \delta\rangle Q\chi . \qquad (\text{XIII.31})$$

Now the function $Q(\rho)\chi(\rho)$ is alternating since it is the product of an alternating function and an invariant one. Since

$$\chi(\rho) = \sum_{M} n_M \exp\langle M, \rho\rangle , \qquad (\text{XIII.32})$$

and

$$Q(\rho) = \sum_{S \in W} (\det S) \exp\langle S\delta, \rho\rangle , \qquad (\text{XIII.33})$$

the product must be of the form

$$Q(\rho)\chi(\rho) = \sigma \sum_{N} c_N \exp\langle N, \rho\rangle , \qquad (\text{XIII.34})$$

where $N$ is of the form $M + S\delta$ where $M$ is a weight and where $S$ is in the Weyl group. Substituting into the differential equation, we see that $M$ contributes only if

$$\langle S^{-1}M + \delta, S^{-1}M + \delta\rangle = \langle \Lambda + \delta, \Lambda + \delta\rangle . \qquad (\text{XIII.35})$$

In fact, we can show that Eq. (XIII.35) is satisfied only when $S^{-1}M = \Lambda$. We first note that $\langle SM + \delta, SM + \delta \rangle$ is maximized for fixed $M$ when $SM$ has only non-negative Dynkin coefficients. Indeed if $h_i = 2h_{\alpha_i}/\langle \alpha_i, \alpha_i \rangle$ and if $M(h_i) < 0$, then $\langle S_i M + \delta, S_i M + \delta \rangle - \langle M + \delta, M + \delta \rangle = \langle -M(h_i)\alpha_i, 2M + 2\delta - M(h_i)\alpha_i \rangle = -\langle \alpha_i, \alpha_i \rangle M(h_i) > 0$. Now consider $M < \Lambda$ with $M(h_i) \geq 0$. Then, by similar arguments, it is easy to show that $\langle \Lambda + \delta, \Lambda + \delta \rangle > \langle M + \delta, M + \delta \rangle$. It follows that the sum in Eq. (XIII.34) need contain only the single term for $N = \Lambda + \delta$. By comparison with the definitions of $Q(\rho)$ and $\chi(\rho)$, it is easy to see that the overall coefficient is unity, so

$$Q(\rho)\chi(\rho) = \sum_{S \in W} (\det S) \exp\langle \Lambda + \delta, S\rho \rangle . \tag{XIII.36}$$

This then yields Weyl's character formula

$$\chi(\rho) = \frac{\sum_{S \in W} (\det S) \exp\langle \Lambda + \delta, S\rho \rangle}{\sum_{S \in W} (\det S) \exp\langle \delta, S\rho \rangle} . \tag{XIII.37}$$

More useful for our purposes is the less general formula which gives the dimension of an irreducible representation. It is clear that this dimension is the value of $\chi(\rho = 0)$. This cannot be obtained simply by setting $\rho = 0$, but must be obtained as a limit. We choose $\rho = t\delta$ and let $t \to 0$. This gives

$$\chi(t\delta) = \frac{\sum_{S \in W} (\det S) \exp\langle S(\Lambda + \delta), t\delta \rangle}{\sum_{S \in W} (\det S) \exp\langle S\delta, t\delta \rangle}$$

$$= \frac{Q(t(\Lambda + \delta))}{Q(t\delta)}$$

$$= \exp\langle -\delta, t\Lambda \rangle \prod_{\alpha > 0} \frac{\exp\langle \alpha, t(\Lambda + \delta) \rangle - 1}{\exp\langle \alpha, t\delta \rangle - 1} . \tag{XIII.38}$$

In this expression we can let $t \to 0$ and find the dimensionality , dim $R = \chi(0)$:

$$\dim R = \prod_{\alpha > 0} \frac{\langle \alpha, \Lambda + \delta \rangle}{\langle \alpha, \delta \rangle} . \tag{XIII.39}$$

To evaluate this expression, we write each positive root, $\alpha$, in terms of the simple roots $\alpha_i$:

$$\alpha = \sum_i k_\alpha^i \alpha_i . \tag{XIII.40}$$

Suppose the Dynkin coefficients of $\Lambda$ are $\Lambda(h_i) = \Lambda_i$, where $h_i = 2h_{\alpha_i}/\langle \alpha_i, \alpha_i \rangle$. Then we have

$$\dim R = \prod_{\alpha > 0} \frac{\sum_i k_\alpha^i (\Lambda_i + 1)\langle \alpha_i, \alpha_i \rangle}{\sum_i k_\alpha^i \langle \alpha_i, \alpha_i \rangle} . \tag{XIII.41}$$

The algebras $A_n, D_n, E_6, E_7$, and $E_8$ have simple roots all of one size, so for them we can drop the factors of $\langle \alpha_i, \alpha_i \rangle$ in Eq. (XIII.41).

Let us illustrate this marvelous formula with a number of examples. Consider first SU(3), that is, $A_2$. The simple roots are all the same size so we ignore the factor $\langle \alpha_i, \alpha_i \rangle$. The positive roots are $\alpha_1, \alpha_2$, and $\alpha_1 + \alpha_2$, which we shall abbreviate here by (1), (2), and (12). Suppose the irreducible representation at hand has a highest weight with Dynkin coefficients $(m_1, m_2)$, then we compute

$$\dim R = \left( \frac{m_1 + 1}{1} \right) \left( \frac{m_2 + 1}{1} \right) \left( \frac{m_1 + m_2 + 2}{2} \right) . \tag{XIII.42}$$

From this example and the fundamental formula, Eq. (XIII.41), we see that the rule for finding the dimensionality of an irreducible representation may be phrased as follows: *The dimension is a product of factors, one for each positive root of the algebra. Each factor has a denominator which is the number of simple roots which compose the positive root. The numerator is a sum over the simple roots in the positive root, with each simple root contributing unity plus the value of the Dynkin coefficient corresponding to the simple root. If the simple roots are not all the same size, each contribution to the numerator and to the denominator must be weighted by $\langle \alpha_i, \alpha_i \rangle$.*

Let us consider a more complicated application of Weyl's formula. The algebra $G_2$ has, as we have seen, fourteen roots, of which six are positive. If the simple roots are denoted $\alpha_1$ and $\alpha_2$ with the latter being the smaller, then the square of $\alpha_1$ is three times larger than that of $\alpha_2$. The positive roots are $\alpha_1, \alpha_2$, $\alpha_1 + \alpha_2, \alpha_1 + 2\alpha_2, \alpha_1 + 3\alpha_2$, and $2\alpha_1 + 3\alpha_2$, which we denote here $(1), (2), (12), (12^2)$, $(12^3)$, and $(1^2 2^3)$. We compute below the dimensions of the representations with the highest weights $(0,1)$ and $(1,0)$, where the first entry pertains to $\alpha_1$ and the second to $\alpha_2$.

| | $(0,1)$ | $(1,0)$ |
|---|---|---|
| $(1), (2)$ | $\frac{3}{3}\,\frac{2}{1}$ | $\frac{3(2)}{3}\,\frac{1}{1}$ |
| $(12)$ | $\frac{3+2}{3+1}$ | $\frac{3(2)+1}{3+1}$ |
| $(12^2)$ | $\frac{3+2+2}{3+1+1}$ | $\frac{3(2)+1+1}{3+1+1}$ |
| $(12^3)$ | $\frac{3+2+2+2}{3+1+1+1}$ | $\frac{3(2)+1+1+1}{3+1+1+1}$ |
| $(1^2 2^3)$ | $\frac{3+3+2+2+2}{3+3+1+1+1}$ | $\frac{3(2)+3(2)+1+1+1}{3+3+1+1+1}$ |
| | dim $= 7$ | dim $= 14$ |

As yet another example, we consider SO(10), that is, $D_5$. The simple roots are numbered so that $\alpha_4$ and $\alpha_5$ are the ones which form the fork at the end of the Dynkin diagram. There are 45 roots of which 20 are positive. Below we calculate the dimensionality of the representations with highest weights $(1,0,0,0,0)$, $(0,0,0,0,1)$, and $(0,0,0,0,2)$.

| | $(1,0,0,0,0)$ | $(0,0,0,0,1)$ | $(0,0,0,0,2)$ |
|---|---|---|---|
| $(1), (2), (3), (4), (5)$ | $2$ | $2$ | $3$ |
| $(12), (23), (34), (35)$ | $\frac{3}{2}$ | $\frac{3}{2}$ | $\frac{4}{2}$ |
| $(123), (234), (235), (345)$ | $\frac{4}{3}$ | $\frac{4}{3}\,\frac{4}{3}$ | $\frac{5}{3}\,\frac{5}{3}$ |
| $(1234), (2345), (1235)$ | $\frac{5}{4}\,\frac{5}{4}$ | $\frac{5}{4}\,\frac{5}{4}$ | $\frac{6}{4}\,\frac{6}{4}$ |
| $(12345), (23^2 45)$ | $\frac{6}{5}$ | $\frac{6}{5}\,\frac{6}{5}$ | $\frac{7}{5}\,\frac{7}{5}$ |
| $(123^2 45)$ | $\frac{7}{6}$ | $\frac{7}{6}$ | $\frac{8}{6}$ |
| $(12^2 3^2 45)$ | $\frac{8}{7}$ | $\frac{8}{7}$ | $\frac{9}{7}$ |
| | dim $R = 10$ | dim $R = 16$ | dim $R = 126$ |

With little effort, we can derive a general formula for the dimensionality of an irreducible representation of SU(n), that is $A_{n-1}$. In the same notation as above, the roots are $(1), (2), \ldots (12), (23), \ldots (123), (234), \ldots, (123 \ldots n)$. We compute the dimensionality for the representation whose highest weight is $(m_1, m_2, \ldots m_n)$:

$(1), (2), \ldots (n)$  $\qquad \frac{m_1+1}{1} \frac{m_2+1}{1} \ldots \frac{m_n+1}{1}$

$(12), (23), \ldots (n-1 \ n)$ $\qquad \frac{m_1+m_2+2}{2} \frac{m_2+m_3+2}{2} \frac{m_{n-1}+m_n+2}{2}$

$\ldots$

$(12 \ldots n)$ $\qquad \frac{m_1+m_2+\ldots m_n+n}{n}$

It is a simple matter to multiply all these factors to find the dimension of the representation.

We can recognize the correspondence between the Dynkin notation and the more familiar Young tableaux if we start with the fundamental representation, $(1, 0, \ldots 0)$, and take the k-times anti-symmetric product of this representation with itself to obtain $(0, 0, \ldots, m_k = 1, 0 \ldots 0)$. This corresponds to the tableau with one column of $k$ boxes. More generally, $(m_1, m_2, \ldots m_n)$ corresponds to the tableau with $m_k$ columns of $k$ boxes.

# References

We follow closely Jacobson's version of the Freudenthal derivation of the Weyl formula, except that we have adopted a less formal language. See JACOBSON,pp. 249–259.

# Exercises

1. Determine the dimensionality of the SO(10) representations $(2, 0, 0, 0, 0)$ and $(0, 1, 0, 0, 0)$.

2. Determine the dimensionality of the $E_6$ representations $(1, 0, 0, 0, 0, 0)$ and $(0, 0, 0, 0, 0, 1)$.

3. Show that the $E_6$ representation $(11, 12, 11, 12, 12, 11)$ has a dimensionality divisible by 137.

# XIV. Reducing Product Representations

In Chapter X we began the consideration of finding the irreducible components of a product representation. The procedure for SU(2) is familiar and trivial. The product of the representations characterized by $j_1$ and $j_2$, the maximal values of $T_z$, contains the irreducible representations for $j$ such that $|j_1 - j_2| \leq j \leq j_1 + j_2$ once each (of course we take only integral $j$ if $j_1 + j_2$ is integral and $j$ half integral otherwise). For SU(n), the reduction is most easily obtained by the method of Young Tableaux. The general solution to the problem of finding the irreducible components of the product of two irreducible representations of a simple Lie algebra can be obtained from the Weyl character formula, but the result (Kostant's formula) involves a double sum over the Weyl group and is not particularly practical. In this chapter, we introduce some techniques which are sufficient for solving the problem in most cases of moderate dimensions.

Of course we already have sufficient tools to solve the problem by brute force. We can calculate all the weights of the representations associated with highest weights $\Lambda_1$ and $\Lambda_2$, and using the Freudenthal recursion relations we can determine the multiplicity of each weight. Then the multiplicity of the weight $M$ in the product representation is found by writing $M = M_1 + M_2$, where $M_1$ and $M_2$ are weights of the two irreducible representations. The multiplicity of the weight $M$ in the product representation is

$$n_M = \sum_{M=M_1+M_2} n_{M_1} n_{M_2} \qquad\qquad \text{(XIV.1)}$$

where $n_{M_1}$ and $n_{M_2}$ are the multiplicities of the weights $M_1$ and $M_2$. Now we know $\Lambda_1 + \Lambda_2$ is the highest weight of one irreducible component so we can subtract its weights (with proper multiplicities) from the list. Now the highest remaining weight must be the highest weight of some irreducible component which again we find and eliminate from the list. Continuing this way, we exhaust the list of weights, and most likely, ourselves.

A more practical approach is to use some relations which restrict the possible choices of irreducible components. The first such relation is the obvious one: $\Lambda_1 + \Lambda_2$ is the highest weight of one irreducible component. The second relation is a generalization of the rule demonstrated before for the anti-symmetric product of a representation with itself.

**Dynkin's Theorem for the Second Highest Representation** provides a simple way to find one or more irreducible representations beyond the highest. Suppose we have two irreducible representations with highest weights $\Lambda_1$ and $\Lambda_2$, with highest weight vectors $\xi_0$ and $\eta_0$. We shall say that two elements of the root space, $\beta$ and $\alpha$ are joined if $\langle \alpha, \beta \rangle \neq 0$. Moreover, we shall say that a chain of simple roots, $\alpha_1, \alpha_2, \ldots \alpha_k$, connects $\Lambda_1$ and $\Lambda_2$ if $\Lambda_1$ is joined to $\alpha_1$ but no other of the $\alpha$'s, $\Lambda_2$ is joined to $\alpha_k$ and no other $\alpha$ in the chain, and each $\alpha_i$ is joined only to the succeeding and preceding $\alpha$'s. We can represent this with a Dynkin diagram by adding a dot for each of $\Lambda_1$ and $\Lambda_2$ and connecting them by segments to the simple roots with which they are joined. Then a chain is the shortest path between the weights $\Lambda_1$ and $\Lambda_2$. For example, consider SU(8) and the representations (1,1,0,0,0,0,0) and (0,0,0,0,2,1,0).

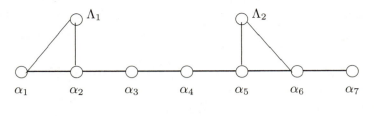

Fig. V.1

Here $\{\alpha_2, \alpha_3, \alpha_4, \alpha_5\}$ is a chain joining $\Lambda_1$ and $\Lambda_2$, but $\{\alpha_1, \alpha_2, \alpha_3, \alpha_4, \alpha_5\}$ is not.

Dynkin's theorem tells us that if $\alpha_1, \ldots \alpha_k$ is a chain joining $\Lambda_1$ and $\Lambda_2$, then $\Lambda_1 + \Lambda_2 - \alpha_1 - \alpha_2 - \ldots - \alpha_k$ is the highest weight of an irreducible representation in the product representation formed from the irreducible representations with highest weights $\Lambda_1$ and $\Lambda_2$. In the above example, the product representation contains $(1,1,0,0,2,1,0) - \alpha_2 - \alpha_3 - \alpha_4 - \alpha_5 = (2,0,0,0,1,2,0)$.

The theorem is proved by establishing that there is a weight vector with the weight described by the theorem which is annihilated by all the raising operators associated with the simple roots. Thus this weight vector generates, through the lowering operators, a separate irreducible representation. Of course, there are other weight vectors with the same weight which do not share this property.

We begin by constructing a sequence of weight vectors starting with $\xi_0$. Since $\langle \Lambda_1, \alpha_1 \rangle \neq 0$ and must be non-negative, it is positive and thus $\Lambda_1 - \alpha_1$ is a weight and has a weight vector

$$\xi_1 = E_{-\alpha_1} \xi_0 \ . \tag{XIV.2}$$

Also, since $\langle \Lambda_1, \alpha_2 \rangle = 0$, $\Lambda_1 - \alpha_2$ is not a weight. However, $\Lambda_1 - \alpha_1 - \alpha_2$ is a weight since $\langle \Lambda_1 - \alpha_1, \alpha_2 \rangle = \langle -\alpha_1, \alpha_2 \rangle > 0$. Proceeding in this way we construct

$$\xi_j = E_{-\alpha_j} \xi_{j-1}$$

$$= E_{-\alpha_j} \ldots E_{-\alpha_1} \xi_0 \ . \tag{XIV.3}$$

Any reordering of the lowering operators in the sequence results in the quantity vanishing just as $E_{-\alpha_2} \xi_0 = 0$.

Now consider

$$E_{\alpha m} \xi_j = E_{\alpha m} E_{-\alpha_j} \xi_{j-1}. \tag{XIV.4}$$

If $m \neq j$, $\left[E_{\alpha m}, E_{-\alpha_j}\right] = 0$ since $\alpha_m - \alpha_j$ is not a root. Thus

$$E_{\alpha m} \xi_j = E_{-\alpha_j} E_{\alpha m} \xi_{j-1} . \tag{XIV.5}$$

If we continue commuting $E_{\alpha m}$ through until it hits $\xi_0$, we get zero since $\xi_0$ corresponds to the highest weight. The only alternative is that $E_{-\alpha m}$ occurs and we take the term

$$E_{-\alpha_j} \ldots [E_{\alpha m}, E_{-\alpha m}] \ldots E_{-\alpha_1} \xi_0 . \tag{XIV.6}$$

But this vanishes since the commutator is just an $H$ which we can move outside (picking up some constant terms) leaving a series of $E_\alpha$'s which are not in the proper order. Thus $E_{\alpha m} \xi_j = 0$ unless $m = j$.

In the event $m = j$, we compute

$$E_{\alpha_j} \xi_j = \left[H_{\alpha_j} + E_{-\alpha_j} E_{\alpha_j}\right] \xi_{j-1}$$

$$= H_{\alpha_j} \xi_{j-1}$$

$$= \langle \Lambda_1 - \alpha_1 \ldots - \alpha_{j-1}, \alpha_j \rangle \xi_{j-1}$$

$$= -\langle \alpha_{j-1}, \alpha_j \rangle \xi_{j-1} \qquad (j > 1)$$

$$E_{\alpha_1} \xi_1 = \langle \Lambda_1, \alpha_1 \rangle \xi_0 . \tag{XIV.7}$$

At the other end of the chain we have an analogous situation. We define

$$E_{-\alpha_k} \eta_0 = \eta_1$$

$$E_{-\alpha_{k-j}} \eta_j = \eta_{j+1} \tag{XIV.8}$$

and find

$$E_{\alpha_m} \eta_{k-j+1} = 0. \qquad (m \neq j),$$

$$E_{\alpha_j} \eta_{k-j+1} = -\langle \alpha_{j+1}, \alpha_j \rangle \eta_{k-j} \quad (j < k)$$

$$E_{\alpha_k} \eta_1 = \langle \Lambda_2, \alpha_k \rangle \eta_0 . \tag{XIV.9}$$

We are now in a position to establish the existence of a vector with weight $\Lambda_1 + \Lambda_2 - \alpha_1 - \ldots - \alpha_k$ which is annihilated by every $E_{\alpha_j}$, for $\alpha_j$ a simple root. This will then be a weight vector for the highest weight of the desired irreducible representation. The most general weight vector with this weight is, using the results of the discussion above,

$$\zeta = \sum_{s=0}^{k} c_s \xi_s \otimes \eta_{k-s} . \tag{XIV.10}$$

We simply choose the coefficients so that the vector is annihilated by every raising operator. For $j \neq 1, k$:

$$E_{\alpha_j} \zeta = c_j E_{\alpha_j} \xi_j \otimes \eta_{k-j} + c_{j-1} \xi_{j-1} \otimes E_{\alpha_j} \eta_{k-j+1}$$

$$= [-c_j \langle \alpha_{j-1}, \alpha_j \rangle - c_{j-1} \langle \alpha_j, \alpha_{j+1} \rangle] \xi_{j-1} \otimes \eta_{k-j}$$

$$= 0 . \tag{XIV.11}$$

Thus for $j \neq 1, k$

$$c_j \langle \alpha_{j-1}, \alpha_j \rangle + c_{j-1} \langle \alpha_j, \alpha_{j+1} \rangle = 0 . \tag{XIV.12}$$

Similarly, considering $j = 1$ and $j = k$,

$$c_1 \langle \Lambda_1, \alpha_1 \rangle - c_0 \langle \alpha_1, \alpha_2 \rangle = 0 ,$$

$$-c_k \langle \alpha_k, \alpha_{k-1} \rangle + c_{k-1} \langle \Lambda_2, \alpha_k \rangle = 0 . \tag{XIV.13}$$

It is clear we can solve these equations, say with $c_0 = 1$. Thus $\zeta$ exists and so does the asserted representation.

The second technique we use to reduce product representations is **Dynkin's method of parts**. This is very easy to apply. If some of the dots in a Dynkin diagram of, say, a simple Lie algebra are deleted, the result is the diagram of a semi-simple subalgebra of the original algebra. If the original diagram was marked with integers to indicate an irreducible representation, the truncated diagram will represent a particular irreducible representation of the subalgebra. Now if we consider two irreducible representations of the original algebra there are associated two irreducible representations of the subalgebra. If we compare the product representations formed by both the representations of the full algebra and those of the subalgebra, it turns out that each irreducible component of the product representation of the subalgebra has a diagram which is a "part" of a diagram of an irreducible component of the product representation of the full algebra in the sense that it is obtained again by deleting the same dots as before.

The utility of this technique lies in the possibility that the subalgebra's products may be easier to deal with than those of the full algebra. In particular, if we delete some dots so that the remaining algebra is in the series $A_n$, the products can be calculated using the well-known technique of Young tableaux. For example, by deleting one dot from $E_6$ we get $A_5$, $D_5$, $A_4 + A_1$, or $A_2 + A_2 + A_1$, each of which is somewhat easier to deal with.

Before proving the correctness of the method of parts, let us consider an example. As a preliminary, note that for $D_5$, i.e. SO(10), the square of the ten-dimensional representation, (1,0,0,0,0) is given by (2,0,0,0,0) + (0,1,0,0,0) + (0,0,0,0,0). This is easy to see because the first piece follows from the rule for the highest weight of a product representation. The second follows for the rule for the second highest weight, or the rule for the anti-symmetric product of an irreducible representation with itself. Use of the Weyl dimension formula reveals that the dimension of the (2,0,0,0,0) representation is 54 and that of the (0,1,0,0,0) is 45, so the remaining representation is one dimensional, i.e. it is (0,0,0,0,0). (Of course, these results can be obtained by more elementary means!) Now let us try to compute the square of the $E_6$ representation (1,0,0,0,0,0). This is the smallest representation of $E_6$, with dimension 27. Again the rules for the highest weight and second highest weight give (2,0,0,0,0,0) and (0,1,0,0,0,0) as irreducible components of the product representation. A computation reveals their dimensions to be 351 (both of them!). Thus the remaining representation is 27 dimensional. Is it (1,0,0,0,0,0) or (0,0,0,0,1,0)? Let us use the method of parts, deleting the fifth simple root to leave us with the diagram for $D_5$. Now we know that (1,0,0,0,0) squared in $D_5$ is (2,0,0,0,0) + (0,1,0,0,0) + (0,0,0,0,0). The method of parts tells us that each of these can be obtained from the irreducible representations in the $E_6$ product by deleting the fifth root. This clearly works for the first two. Moreover, we see that we must choose (0,0,0,0,1,0) as the final representation of $E_6$.

We proceed to a more formal consideration of the method of parts. Let $G$ be a simple Lie algebra with a basis of simple roots $\{\alpha_i\}$. Select a subset, $\{\beta_j\} \subset \{\alpha_i\}$ and let $G'$ be the semi-simple algebra which is generated by the $e_{\beta_j}$'s and $e_{-\beta_j}$'s. The Cartan subalgebra, $H'$ of $G'$ is contained in the Cartan subalgebra, $H$ of $G$. The Dynkin diagram for $G'$ is obtained from that of $G$ by deleting the appropriate dots.

Suppose $M$ is a weight of a representation of $G$ and $\phi_M$ is an associated weight vector:

$$H\phi_M = M(h)\phi_M . \tag{XIV.14}$$

Then $\phi_M$ is a weight vector for the induced representation of $G'$, since if $h' \in H'$,

$$H'\phi_M = M(h')\phi_M . \tag{XIV.15}$$

Now the weight in the induced representation, $\overline{M}$, has the property $\overline{M}(h') = M(h')$ but differs from $M$ because it can be expressed entirely in terms of the $\beta_j$'s. If we write

$$M = [M - \sum_{i,j} \beta_i \langle M, \beta_j \rangle \langle \beta_j, \beta_i \rangle^{-1}]$$

$$+ \sum_{i,j} \beta_i \langle M, \beta_j \rangle \langle \beta_j, \beta_i \rangle^{-1}$$

$$. \tag{XIV.16}$$

we see that the second term is precisely $\overline{M}$. This is so because every $h'$ is a linear combination of $h_\beta$'s and because the first term vanishes on every $h_\beta$.

We see from Eq. (XIV.16) that $\overline{M}$ has exactly the same Dynkin coefficients with respect to the $\beta$'s as $M$ itself. The Dynkin coefficients with respect to $G'$ are obtained simply by dropping the requisite coefficients from those of $M$. In particular, if $\Lambda$ is the highest weight of a representation of $G$, $\overline{\Lambda}$ is a highest weight of one of the irreducible representations contained in the representation of $G'$ arising from the representation of $G$. For example, the (1,1) representation of $A_2$ (SU(3)) contains the (1) representation of $A_1$ (SU(2)) when we obtain the SU(2) by deleting one dot from the Dynkin diagram of SU(3).

The representation associated with $\overline{\Lambda}$ is obtained by operating on the weight vector for $\Lambda$ with all the lowering operators $E_{\alpha_i}, \alpha_i \in \Sigma'$. It is easy to see that for all vectors $\phi$ in this representation, if $\alpha_j \notin \Sigma'$ then $E_{\alpha_j}\phi = 0$ .

Now consider two irreducible representations, $R_{\Lambda_1}$ and $R_{\Lambda_2}$ of G, where $\Lambda_1$ and $\Lambda_2$ are their highest weights. Associated with these are two irreducible representations of $G'$, $R_{\Lambda_1'}$ and $R_{\Lambda_2'}$. Now consider the product representations $R_{\Lambda_1} \times R_{\Lambda_2}$ and $R_{\Lambda_1'} \times R_{\Lambda_2'}$. In general these are both reducible:

$$R_{\Lambda_1} \times R_{\Lambda_2} = R_{\Lambda_a} + R_{\Lambda_b} + \dots \qquad \text{(XIV.15)}$$

and

$$R_{\Lambda_1'} \times R_{\Lambda_2'} = R_{\Lambda_a'} + R_{\Lambda_b'} + \dots \,. \qquad \text{(XIV.16)}$$

We want to show that for some $\Lambda_a, \overline{\Lambda_a} = \Lambda_a'$, etc. Now consider the highest weight vector of $R_{\Lambda_a'}$. It is annihilated by $E_{\alpha_i}, \alpha_i \in \Sigma'$ and also by all the $E_{\alpha_j}, \alpha_j \in \Sigma$. Thus it is a highest weight also for $G$ as well as for $G'$, and thus defines one of the $R_{\Lambda_a}$'s. Thus every scheme of $R_{\Lambda_1'} \times R_{\Lambda_2'}$ corresponds to one of the schemes of $R_{\Lambda_1} \times R_{\Lambda_2}$ with the appropriate dots deleted.

In reducing product representations for SU(2), it is clear that the product of irreducible representations contains only either integral spin or half-integral spin representations. In Dynkin language, the Cartan matrix is the number 2. The

Dynkin coefficient of the highest weight is a single integer. The lower weights are obtained by subtracting the 2 from the highest weight. Thus if the coefficient of the highest weight is odd, the coefficient of every weight is odd. The oddness or evenness of the the two irreducible representations thus determines the oddness or evenness of all the irreducible representations in their product.

The analogous concept for SU(3) is **triality**. The fundamental representation is said to have triality +1. Every weight in this representation is obtained from (1,0) by subtracting a row of the Cartan matrix. Now consider the quantity $1A_{i1} + 2A_{i2}$. For i=1, this is zero, while for i = 2, it is three. Thus if we calculate for any weight $(a_1, a_2)$ of an irreducible representation the quantity $a_1 + 2a_2 \pmod 3$, it must be the same as it is for the highest weight. Thus for the three dimensional representation we have (1,0), (-1,1), and (0,-1), where $a_1 + 2a_2$ is 1,1, and -2. It is clear that the triality, $a_1 + 2a_2 \pmod 3$, of a representation in the product of two irreducible representations is the sum of the trialities (mod 3) of the components.

If we look at the Cartan matrix for $A_n$, we see that $\sum_j jA_{ij} = 0 \pmod{n+1}$. Thus each irreducible representation of $A_n$ can be characterized by a number $C \equiv \sum_j ja_j \pmod{n+1}$ where the highest weight of the representation has coefficients $(a_1, a_2, \ldots a_n)$. We see that every weight in the irreducible representation will have the same value for $C$. For a product of representations, $R_1$ and $R_2$, each irreducible component has the value $C \equiv C(R_1) + C(R_2)$. For example, consider $5^* \times 5^*$ in SU(5), that is $(0,0,0,1) \times (0,0,0,1) = (0,0,0,2) + (0,0,1,0)$. We have $C((0,0,0,1)) = 4, C((0,0,0,2)) = 8 \equiv 3 \pmod 5, C((0,0,1,0)) = 3$. We refer to the irreducible representations with a fixed value of $C$ as **conjugacy classes**.

For $B_n$, there are two conjugacy classes which are given by $C = a_n \pmod 2$, since the last column of the Cartan matrix is given entirely by even integers. If $C = 1$, the representation is a spinor representation. This nomenclature makes sense even for $B_1$ which is the complex form of O(3). The spinor representations are seen to be the half-integral angular momentum representations.

For the algebras $C_n$, we take $C = a_1 + a_3 + a_5 + \ldots \pmod 2$. We can see that this will work by taking the sum of the first, third, fifth, *etc.* elements in a row of the Cartan matrix and noticing that its value is always even.

The algebra $D_n$ is slightly more complicated. First we notice that for any row of the Cartan matrix, the sum of the last two entries is an even number. Thus we can take $C_1 = a_{n-1} + a_n \pmod 2$. Next consider the case n even. If we take the sum of the first, third, fifth,... elements of a row of the Cartan matrix, it is always even. Thus for n even, we choose $C_2 = a_1 + a_3 + \ldots a_{n-1} \pmod 2$. For n odd, we take $C_2 = a_1 + a_3 + \ldots a_{n-2} + (a_{n-1} - a_n)/2 \pmod 2$.

As an example, consider the SO(10) decomposition of $(0,0,0,1,0) \times (0,0,0,1,0)$, i.e. $16 \times 16$. The 16 has $C_1 = 1$ and $C_2 = \frac{1}{2}$. Thus the irreducible representations in the product will have $C_1 = 2 \equiv 0 \pmod 2$ and $C_2 = 1$. The actual decomposition is $(0,0,0,2,0) + (0,0,1,0,0) + (1,0,0,0,0)$, each piece of which has the proper values of $C_1$ and $C_2$.

An examination of the Cartan matrix for $E_6$ reveals that $A_{i1} + A_{i4} = A_{i2} + A_{i5} \pmod 3$, so we can take $C = a_1 - a_2 + a_4 - a_5 \pmod 3$. Similarly, for $E_7$, we see that $C = a_4 + a_6 + a_7 \pmod 2$ determines the conjugacy classes. The algebras $G_2, F_4$, and $E_8$ have only a single conjugacy class.

As an example of the techniques discussed in this chapter, let us consider the decomposition of $(2,0,0,0,0,0) \times (2,0,0,0,0,0) = 351 \times 351$ in $E_6$. The $(2,0,0,0,0,0)$ is in the conjugacy class $C \equiv 2 \pmod 3$, so all the components in the product will be in the conjugacy class $C \equiv 4 \equiv 1 \pmod 3$. Clearly one irreducible component has highest weight $(4,0,0,0,0,0)$. Using the rule for the second highest weight we find that $(4,0,0,0,0,0) - (2,-1,0,0,0,0) = (2,1,0,0,0,0)$ is a highest weight of an irreducible component. Next we use the method of parts, striking the sixth root to reduce $E_6$ to $A_5$. The products in $A_5$ may be calculated by Young tableaux, with the result $(2,0,0,0,0) \times (2,0,0,0,0) = (4,0,0,0,0) + (2,1,0,0,0) + (0,2,0,0,0)$. Thus the $E_6$ product contains $(4,0,0,0,0,0) + (2,1,0,0,0,0) + (0,2,0,0,0,X)$ where X is a non-negative integer. Next, let us use the parts method, striking the fifth root to reduce $E_6$ to $D_5$. Now we must calculate $(2,0,0,0,0) \times (2,0,0,0,0) = 54 \times 54$ in $D_5$.

This subsidiary calculation is itself a useful example. The 54 in $D_5$ has $C_1 = 0$ and $C_2 = 0$, so the irreducible components of the product must have these values as well. Certainly the $D_5$ product contains $(4,0,0,0,0) + (2,1,0,0,0)$ as we see using the highest weight and second highest weight procedures. Using the parts method to reduce $D_5$ to $A_4$ we see that the product must contain a term $(0,2,0,0,W)$. Since $C_1 \equiv 0$, W is even. It is a fair bet that $(0,2,0,0,2)$ has too high a dimension, so we guess that (0,2,0,0,0) is in the product. Using the Weyl dimension formula, we find the $D_5$ values, dim(4,0,0,0,0)=660, dim(2,1,0,0,0)=1386, dim(0,2,0,0,0)=770. This totals to 2816 so we need an additional 2916-2816=100. The smallest representations in the proper conjugacy class are (0,0,0,0,0), (2,0,0,0,0), (0,1,0,0,0), and (0,0,0,1,1) with dimensions 1, 54, 45, and 210 respectively. Thus we conclude that in $D_5$, $(2,0,0,0,0) \times (2,0,0,0,0) = (4,0,0,0,0) + (2,1,0,0,0) + (0,2,0,0,0) + (2,0,0,0,0) + (0,1,0,0,0) + (0,0,0,0,0)$.

Returning to $E_6$, we note that the representation identified as (0,2,0,0,0,X) must be (0,2,0,0,0,0) in order to account for the $D_5$ representation (0,2,0,0,0). Again, by comparison with the $D_5$ representations, we know that the $E_6$ product must contain, at a minimum, the representations (2,0,0,0,T,0), (0,1,0,0,Y,0), and (0,0,0,0,Z,0), where T,Y, and Z are non-negative integers. We can determine these integers by considering the conjugacy classes. We have $2 - T \equiv 1 \pmod 3$, $-1 - Y \equiv 1 \pmod 3$, and $-Z \equiv 1 \pmod 3$. The smallest solutions are T=1, Y=1, and Z=2. Thus we guess that the $E_6$ decomposition is $(2,0,0,0,0,0) \times (2,0,0,0,0,0) = (4,0,0,0,0,0) + (2,1,0,0,0,0) + (0,2,0,0,0,0) + (2,0,0,0,1,0) + (0,1,0,0,1,0) + (0,0,0,0,2,0)$. Using the Weyl formula, the dimensions of these are determined to be $351 \times 351 = 19,305 + 54,054 + 34,398 + 7,722 + 7,371 + 351 = 123,201$.

This example shows that the hardest work required in reducing such products is simply the evaluation of the Weyl formula for the dimension.

One additional technique for reducing product representations is worth mentioning. We recall from Chapter XI the definition of the index of a representation, $\phi$:

$$\text{Tr } \phi(x)\phi(y) = l_\phi(x,y)_2 \qquad \text{(XIV.19)}$$

where $(\ ,\ )_2$ is proportional to the Killing form but normalized so that the largest roots has a length squared of two.

Now suppose we have a representation which is the sum of two representations, $\phi_1$ and $\phi_2$. Then, clearly

$$l_{\phi_1 + \phi_2} = l_{\phi_1} + l_{\phi_2} . \qquad \text{(XIV.20)}$$

On the other hand, for a product representation, we see that

$$l_{\phi_1 \otimes \phi_2} = N_{\phi_1} l_{\phi_2} + N_{\phi_2} l_{\phi_1} \qquad \text{(XIV.21)}$$

where $N_{\phi_1}$ and $N_{\phi_2}$ are the dimensions of the representations. Since we know how to compute the indices in terms of the Casimir operators, this can be used to reduce significantly the possibilities for the irreducible components of the product representation.

## References

Most of this material is due to E. B. Dynkin. For the method of the second highest representation, see

DYNKIN III, especially pp. 266–268.

For the method of parts see

DYNKIN III, especially pp. 275–280.

For conjugacy classes, see

DYNKIN II, especially pp. 136–137.

Some of the same material is discussed by SLANSKY.

## Exercises

1. Reduce $16 \times 16$ [$16 = (0,0,0,0,1)$] in SO(10).

2. Reduce $14 \times 14$ [$14 = (1,0)$] in $G_2$. Check using indices.

3. Reduce $27 \times 27'$ [for $27 = (1,0,0,0,0,0)$, $27' = (0,0,0,0,1,0)$] in $E_6$.

# XV. Subalgebras

Some examples may be illuminating as an introduction to the topic of subalgebras. Suppose we start with the algebra $G = A_5$, i.e. SU(6), the traceless $6 \times 6$ matrices. Now one subalgebra is obtained by considering only those matrices with non-zero $4 \times 4$ and $2 \times 2$ diagonal pieces, and zeros in the $4 \times 2$ off- diagonal pieces. If the two diagonal blocks are required to be traceless separately, then the restricted set is the subalgebra $G' = A_3 + A_1 \subset A_5 = G$. It is clear that we can take as the Cartan subalgebra $H' \subset G'$ the diagonal matrices, so $H' \subset H$. The dimension of $H'$ is one fewer than that of $H$ since there is a one dimensional subspace of $H$ proportional to the diagonal element which is $+1$ for the first four components and $-2$ on the last two.

The root vectors of $G'$ are just the root vectors of $G$ which have non-zero components only in the two diagonal blocks. If the space proportional to $e_\alpha$ is denoted $G_\alpha$, we have

$$G = H + \sum_{\alpha \in \Sigma} G_\alpha \qquad\qquad\text{(XV.1)}$$

while for some set $\Sigma' \subset \Sigma$

$$G' = H' + \sum_{\alpha \in \Sigma'} G_\alpha . \qquad\qquad\text{(XV.2)}$$

A subalgebra with this property is called **regular**. In addition to this SU(6) example, the subalgebras we employed in the method of parts – which were obtained by deleting dots from Dynkin diagrams – were regular. Not all subalgebras, however, are regular.

Let us consider again $A_5$ and a particular embedding of $G' = A_2 + A_1 (SU(3) \times SU(2))$. We know that every matrix in the Lie algebra of SU(2) is a linear combination of $\sigma_1, \sigma_2,$ and $\sigma_3$ and every matrix in the Lie algebra of SU(3) is a linear combination of $\lambda_1, \lambda_2, \ldots \lambda_8$. Let us add to these $\sigma_0$ and $\lambda_0$ which are the $3 \times 3$ and $2 \times 2$ identity matrices. Now every $6 \times 6$ matrix can be written in terms of

$$\begin{bmatrix} \lambda_i & 0 \\ 0 & \lambda_i \end{bmatrix} , \quad \begin{bmatrix} \lambda_i & 0 \\ 0 & -\lambda_i \end{bmatrix} , \quad \begin{bmatrix} 0 & \lambda_i \\ \lambda_i & 0 \end{bmatrix} , \quad \begin{bmatrix} 0 & -i\lambda_i \\ i\lambda_i & 0 \end{bmatrix} ,$$

i.e. $\sigma_0 \otimes \lambda_i, \sigma_3 \otimes \lambda_i, \sigma_1 \otimes \lambda_i, \sigma_2 \otimes \lambda_i, \quad i = 0, 1, \ldots 8$. Now this is equivalent to regarding the six dimensional vectors in the carrier space as having two indices, with the $\sigma$ acting on the first and the $\lambda$ on the second.

Now suppose we consider only elements of the forms $\sigma_0 \otimes \lambda_i$ and $\sigma_i \otimes \lambda_0$. Then an element of one form commutes with an element of the other. Thus these elements form a subalgebra which is $A_2 + A_1$. The Cartan subalgebra of the $A_2 + A_1$ has a basis $\sigma_3 \otimes \lambda_0, \sigma_0 \otimes \lambda_3, \sigma_0 \otimes \lambda_8$. The root vectors are $\sigma_+ \otimes \lambda_0, \sigma_- \otimes \lambda_0, \sigma_0 \otimes t_+, \sigma_0 \otimes t_-, \sigma_0 \otimes u_+, \sigma_0 \otimes u_-, \sigma_0 \otimes v_+,$ and $\sigma_0 \otimes v_-$. We see that $H' \subset H$. However, the root vectors of $G'$ are not among those of $G$. Thus, for example,

$$\sigma_+ \otimes \lambda_0 = \begin{bmatrix} 0 & \lambda_0 \\ 0 & 0 \end{bmatrix} . \qquad\qquad\text{(XV.3)}$$

We cannot write $G'$ in the form Eq. (XV.2), so the subalgebra is not regular.

The six dimensional representation of $A_5$ gave a reducible representation of the regular subalgebra, $A_3 + A_1$: $6 \to (4, 1) + (1, 2)$. The non-regular subalgebra, $A_2 + A_1$, gave an irreducible representation: $6 \to (3, 2)$. As we shall see, this is typical.

As a further example of regular and non-regular subalgebras, consider SU(2) as a subalgebra of SU(3). If the SU(2) is generated by $t_+, t_-$, and $t_z$, the SU(2) is a regular subalgebra. On the other hand, there is a three dimensional representation of SU(2). The $3 \times 3$ matrices of this representation are elements of SU(3) so this provides a second embedding of SU(2) in SU(3), which is not regular. Under the regular embedding, the 3 dimensional representation of SU(3) becomes a reducible $2 + 1$ dimensional representation of SU(2), while under the second embedding, it becomes an irreducible representation of SU(2).

It is clear that a moderate sized algebra may have an enormous number of subalgebras. To organize the task of finding them we introduce the concept of a **maximal subalgebra**. $G'$ is a maximal subalgebra of $G$ if there is no larger subalgebra containing it except $G$ itself. Now we can proceed in stages finding the maximal subalgebras, then their maximal subalgebras, and so on.

There is a slight flaw in this approach. A maximal subalgebra of a semi-simple algebra need not be semi-simple. Consider, for example, SU(2) and the subalgebra generated by $t_+$ and $t_z$. It is certainly maximal, since if we enlarge it we shall have all of SU(2). However, the subalgebra is not simple because $t_+$ generates an ideal in it. We shall generally restrict ourselves to the consideration of maximal semi-simple subalgebras, that is, semi-simple algebras contained in no other semi-simple subalgebras except the full algebra.

Dynkin introduced the notions of an **R-subalgebra** and an **S-subalgebra**. An R-subalgebra is a subalgebra which is contained in some regular subalgebra. An S-subalgebra is one which is not. The task then is to find the regular maximal subalgebras and the maximal S-subalgebras. The regular subalgebras are more easily dealt with.

Suppose $G'$ is a regular subalgebra of a simple algebra $G$, $\Sigma' \subset \Sigma$ is the set of its roots, and $\Pi' \subset \Sigma'$ is a basis of simple roots for $G'$. Now if $\alpha', \beta' \in \Pi'$, then $\alpha' - \beta' \notin \Sigma'$. In fact $\alpha' - \beta' \notin \Sigma$, since if $\alpha' - \beta' \in \Sigma$, $[e_{\alpha'}, e_{-\beta'}] \sim e_{\alpha' - \beta'}$ so then $\alpha' - \beta' \in \Sigma'$. Thus to find regular subalgebras of $G$ we seek sets $\Pi' \subset \Sigma$ such that $\alpha', \beta' \in \Pi' \Rightarrow \alpha' - \beta' \notin \Sigma$. Then we take as $G'$ the subalgebra generated by the $e_{\alpha'}, e_{-\alpha'}, h_{\alpha'} \in \Pi'$.

An algorithm for this has been provided by Dynkin. Start with $\Pi$, the simple roots of $G$. Enlarge it to $\overline{\Pi}$ by adding the most negative root in $\Sigma$. Now $\overline{\Pi}$ has the property that if $\alpha', \beta' \in \overline{\Pi}$, then $\alpha' - \beta' \notin \Sigma$. However, $\overline{\Pi}$ is a linearly dependent set. Thus, if we eliminate one or more vectors from $\overline{\Pi}$ to form $\Pi'$, it will have the required properties. In general, $\Pi'$ will generate a semi-simple algebra, not a simple one.

This procedure is easy to follow using Dynkin diagrams. We form the **extended Dynkin diagram** associated with $\overline{\Pi}$ by noting that the vector added to $\Pi$ is the negative of the highest weight ,$\gamma$, of the adjoint representation. Since we know the Dynkin coefficients of this weight, it is easy to add the appropriate dot. For $B_n$ the adjoint representation has highest weight $(0, 1, 0, \ldots 0)$, so the extended diagram is

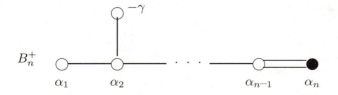

Similarly, for $D_n$, the highest weight of the adjoint is $(0, 1, 0, \ldots 0)$ so the extended diagram is

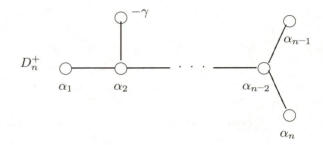

In an analogous fashion, we find the remaining extended diagrams:

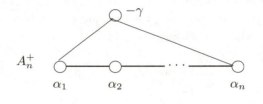

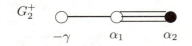

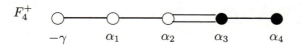

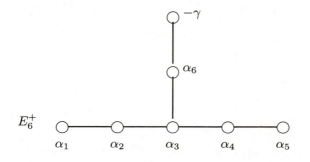

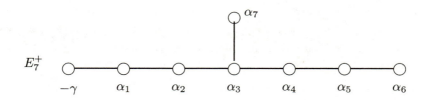

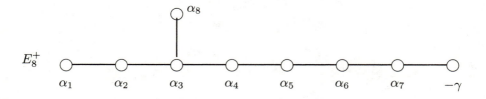

A few examples will make clear the application of these diagrams. By deleting a single dot from the extended diagram for $G_2$ we obtain in addition to the diagram for $G_2$ itself

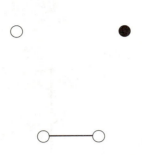

Thus we see that $A_2$ and $A_1 + A_1$ are regular subalgebras of $G_2$. Starting with $B_6$ we find among others, the subalgebra $B_3 + D_3$. In other words, we have $O(13) \supset O(7) \times O(6)$.

The $A_n$ algebras are somewhat pathological. If we remove a single dot from the extended diagram, the result is simply the original diagram. If we remove two dots we obtain a regular subalgebra, but one that is maximal only among the semi-simple subalgebras, not maximal among all the subalgebras. This is actually familiar: from SU(5) one obtains not just SU(3) x SU(2), but also SU(3) x SU(2) x U(1), which itself lies in a larger, non-semi-simple subalgebra of SU(5).

Dynkin's rule[1] for finding the maximal regular subalgebras is this: the regular semi-simple maximal subalgebras are obtained by removing one dot from the extended Dynkin diagram. The non-semi-simple maximal regular subalgebras are obtained by selecting one of the simple roots, $\alpha \in \Pi$, and finding the subalgebra generated by $e_\alpha$ and $h_\alpha$, together with $e_\beta, e_{-\beta},$ and $h_\beta$ for all the simple roots $\beta$ other than $\alpha$. Such a non-semi-simple algebra contains a semi-simple subalgebra generated by excluding $e_\alpha$ and $h_\alpha$ as well. This may be maximal among the semi-simple subalgebras, or it may be contained in an S-subalgebra.

Dynkin's rule has been shown to be not entirely correct.[2] In particular, it would have $A_3 + A_1$ maximal in $F_4$ while in fact $A_3 + A_1 \subset B_4 \subset F_4$. Similarly, $A_3 + A_3 + A_1 \subset D_6 + A_1 \subset E_7$ and in $E_8$, $A_3 + D_5 \subset D_8$, $A_1 + A_2 + A_5 \subset$

$A_2 + E_6$, and $A_7 + A_1 \subset E_7 + A_1$.

In an analogy with the example $SU(4) \times SU(2) \subset SU(6)$, we see that $SU(s) \times SU(t) \subset SU(s+t), O(s) \times O(t) \subset 0(s+t)$, and $Sp(2s) \times Sp(2t) \subset Sp(2s+2t)$, where the embedding is the obvious one. Each of these gives a regular subalgebra except for $O(s) \times O(t)$ when $s$ and $t$ are odd, as is easily verified from the extended diagrams. The last embedding is thus an S-subalgebra.

We have already seen two examples of maximal S-subalgebras. One was the embedding $A_2 + A_1 \subset A_5$ which produced the decomposition of the fundamental representation $6 \to (3, 2)$. We can simplify the notation in this section by passing from the Lie algebras to the associated groups. Thus we have $SU(6) \supset SU(3) \times SU(2)$. More generally, we have $SU(st) \supset SU(s) \times SU(t)$ as a (non-simple) maximal S-subalgebra.

For the orthogonal groups we follow the path used in Chapter VIII. Rather than require $A^t A = I$, we consider the more general relation $B^t K B = K$ where $K$ is a symmetric $n \times n$ matrix. This is the same as taking all $n \times n$ matrices, $B$, which preserve a symmetric bilinear form $(\phi, \eta) = \sum_{i,j} \phi_i K_{ij} \eta_j, K_{ij} = K_{ji}$. Now consider the groups $O(s_1)$ and $O(s_2)$ preserving $(\phi_1, \eta_1)_1$ and $(\phi_2, \eta_2)_2$. If we consider the $s_1 s_2$ dimensional space spanned by vectors like $\phi_1 \otimes \phi_2$, we have a symmetric bilinear form defined by $(\phi_1 \otimes \phi_2, \eta_1 \otimes \eta_2) = (\phi_1, \eta_1)_1 (\phi_2, \eta_2)_2$. The subgroup $O(s_1) \times O(s_2)$ acts as $B_1 \otimes B_2(\phi_1 \otimes \phi_2) = B_1 \phi_1 \otimes B_2 \phi_2$ for $B_1 \in O(s_1)$ and $B_2 \in O(s_2)$. It is clear that this subgroup indeed leaves the symmetric bilinear form invariant and thus $O(s_1) \times O(s_2) \subset O(s_1 s_2)$. Indeed, it is a maximal S-subgroup.

In a similar fashion, we can consider $Sp(2n)$ to be the set of $2n \times 2n$ matrices preserving an anti-symmetric bilinear form: $(\phi, \eta) = -(\eta, \phi)$. Now if we take $Sp(2n_1) \times Sp(2n_2)$ we will act on a space of dimension $4n_1 n_2$. With $(\phi_1 \otimes \phi_2, \eta_1 \otimes \eta_2) = (\phi_1, \eta_1)_1 (\phi_2, \eta_2)_2$ we see that the form is symmetric and $Sp(2n_1) \times Sp(2n_2) \subset O(4n_1 n_2)$. Analogously, $Sp(2n) \times O(s) \subset Sp(2ns)$ as a maximal S-subgroup.

The other maximal S-subalgebra we have encountered is in the embedding $SU(2) \subset SU(3)$ whereby the three dimensional representation of $SU(3)$ becomes the three dimensional representation of $SU(2)$. Since $SU(2)$ has a three dimensional representation by $3 \times 3$ matrices, it is bound to be a subalgebra of $SU(3)$. More generally, if $G$ is a simple algebra with an $n$-dimensional representation, $G \subset SU(n)$. Is $G$ then maximal in $SU(n)$? For the most part, the answer is this: if the $n$-dimensional representation of $G$ has an invariant symmetric bilinear form, $G$ is maximal in $SO(n)$, if it has an invariant anti-symmetric bilinear form it is maximal in $Sp(n)$ ($n$ must be even). If it has no invariant bilinear form, it is maximal in $SU(n)$. The exceptions to this are few in number and have been detailed by Dynkin.[3]

Let us consider an example with $SU(3)$, which has an eight dimensional adjoint representation. Rather than think of the vectors in the eight dimensional space as columns, it is convenient to think of them as $3 \times 3$ traceless matrices:

$$\phi = \sum_i \phi_i \lambda_i \qquad (\text{XV.4})$$

where the $\lambda_i$ are those of Eq. (II.1). Now we remember that the adjoint representation is given by

$$ad\, x \lambda_i = [x, \lambda_i] \ . \qquad (\text{XV.5})$$

The invariance of a symmetric form $(\phi, \eta)$ under an infinitesimal transformation $\exp \mathcal{B} \approx I + \mathcal{B}$ yields

$$(\mathcal{B}\phi, \eta) + (\phi, \mathcal{B}\eta) = 0 \ . \qquad (\text{XV.6})$$

For the adjoint representation, the linear tranformation $\mathcal{B}$ corresponding to an element $x$ of the Lie algebra is simply $\mathcal{B}\phi = [\mathcal{B}, \phi]$. Now if we define

$$(\phi, \eta) = Tr\ \phi\eta \qquad (\text{XV.7})$$

the invariance of the form follows from the identity

$$Tr\ [x, \phi]\, \eta + Tr\ \phi\, [x, \eta] = 0 \ . \qquad (\text{XV.8})$$

Thus we see there is a symmetric invariant bilinear form for the adjoint representation and thus $SU(3) \subset SO(8)$. Of course the demonstration is more general: $SU(n) \subset SO(n^2 - 1)$.

It is clear that we must learn how to determine when an $n$-dimensional representation of a simple Lie algebra admits a bilinear form and whether the form is symmetric or anti-symmetric. As a beginning, let us consider $SU(2)$ and in particular the $2j + 1$ dimensional representation. We shall construct explicitly the bilinear invariant. Let

$$\xi = \sum_i c_i \phi_i$$

$$\eta = \sum_i d_i \phi_i \qquad \text{(XV.9)}$$

where the $\phi_i$ are a basis for the representation space and

$$T_z \phi_m = m \phi_m$$

$$T_+ \phi_m = \sqrt{j(j+1) - m(m+1)} \phi_{m+1}$$

$$T_- \phi_m = \sqrt{j(j+1) - m(m-1)} \phi_{m-1} . \qquad \text{(XV.10)}$$

Suppose the bilinear form is

$$(\xi, \eta) = \sum_{m,n} a_{mn} c_m d_n . \qquad \text{(XV.11)}$$

Now the invariance of the form requires in particular

$$(T_z \xi, \eta) + (\xi, T_z \eta) = 0 . \qquad \text{(XV.12)}$$

Thus in the sum we must have $m + n = 0$ so

$$(\xi, \eta) = \sum_m a_m c_m d_{-m} . \qquad \text{(XV.13)}$$

If we next consider the requirement

$$(T_-\xi, \eta) + (\xi, T_-\eta) = 0 , \tag{XV.14}$$

we find that up to a multiplicative constant the bilinear form must be

$$(\xi, \eta) = \sum_m (-1)^m c_m d_{-m} . \tag{XV.15}$$

We see that we have been able to construct a bilinear invariant and moreover, if $j$ is integral, the form is symmetric under interchange of $\xi$ and $\eta$, while if $j$ is half-integral it is anti-symmetric.

The generalization to all the simple Lie algebras is only slightly more complicated. From the analogs of Eq. (XV.12), we conclude that we can form a bilinear invariant only if for every weight $M$ in the representation, $-M$ is a weight also. To determine which irreducible representations have this property it suffices to consider the representations from which we can build all others. For example, for $A_n$, the representation with highest weight $(1, 0, \ldots 0)$ does not contain the weight $(-1, 0, \ldots 0)$ but instead $(0, 0, \ldots, 0, -1)$. On the other hand, the adjoint representation whose highest weight is $(1, 0, \ldots 0, 1)$ does contain the weight $(-1, 0, \ldots 0, -1)$. More generally for $A_n$, if the highest weight has the symmetric configuration $(n_1, n_2, \ldots n_2, n_1)$, the weights do occur in opposite pairs, $M$ and $-M$.

The general result is the following. Representations of $B_n, C_n, D_{2n}, G_2, F_4,$ $E_7,$ and $E_8$ always have an invariant bilinear form. The algebras $A_n, D_{2n+1},$ and $E_6$ have invariant bilinear forms for representations of the forms:

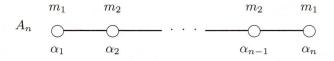

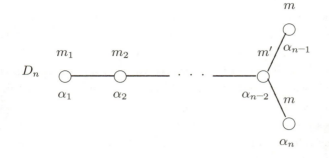

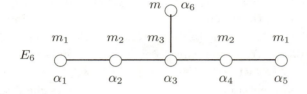

Now as for the symmetry or anti-symmetry of the bilinear form it turns out to be determined by the number of levels in the weight diagram. Dynkin calls the number of levels less one the **height of the representation**. Thus for $SU(2)$, the $2j + 1$ dimensional representation has a height $2j$. For all the simple Lie algebras, just as for SU(2), if the height is even, the form is symmetric, and if the height is odd, the form is anti-symmetric. Now Dynkin has determined the heights of the irreducible representations of the simple Lie algebras in terms of their Dynkin coefficients.[4] The results are summarized in the following Table:

$A_n$      $(n, (n-1)2, \ldots, n)$

$B_n$      $(1 \cdot 2n, 2 \cdot (2n-1), \ldots, (n-1)(n+2), n(n+1)/2)$

$C_n$      $\left(1 \cdot (2n-2), 2(2n-2), \ldots, (n-1)(n+1), n^2\right)$

$D_n$      $(1 \cdot (2n-2), 2 \cdot (2n-3), \ldots, (n-2)(n+1), n(n-1)/2, n(n-1)/2)$

$G_2$      $(10, 6)$

$F_4$      $(22, 42, 30, 16)$

$E_6$      $(16, 30, 42, 30, 16, 22)$

$E_7$      $(34, 66, 96, 75, 52, 27, 49)$

$E_8$      $(92, 182, 270, 220, 168, 114, 58, 136)$

The height of the representation is calculated by multiplying each Dynkin coefficient by the number from the table corresponding to its location. Thus, for example, the adjoint representation of $A_n$, with Dynkin coefficients $(1, 0, \ldots 0, 1)$ has height $2n$. It is clear that to determine whether a bilinear form is symmetric or anti-symmetric we need only consider those entries in the table and those Dynkin coefficients which are odd. It is apparent that since $SU(3)$ representations have a bilinear form only if the Dynkin coefficients of the highest weight are of the form $(n, n)$, all such bilinear forms are symmetric. On the other hand, we see that $SU(6)$ has a representation with highest weight $(0, 0, 1, 0, 0)$ and dimension 20 which has an anti-symmetric bilinear invariant. Thus $SU(6) \subset Sp(20)$.

There are a few instances in which the procedure described above does not identify a maximal subalgebra. These exceptions have been listed by Dynkin[5] and we shall not dwell on them here. One example will suffice to indicate the nature of these exceptions. There is a 15 dimensional representation of $A_2$, which has a highest weight $(2, 1)$. We would expect this to be maximal in $A_{14}$ since it has no bilinear invariant. In fact, there is an embedding of $A_2$ in $A_5$ under which the 15 dimensional representation of $A_5$, $(0, 1, 0, 0, 0)$, becomes the fifteen dimensional representation of $A_2$. Thus we have the chain $A_2 \subset A_5 \subset A_{14}$. We can understand the embedding of $A_2$ in $A_5$ as follows. Since $A_2$ has a six dimensional representation, it is maximal in $A_5$, i.e. $SU(6)$. The anti-symmetric product of the six with itself, in both $A_2$ and $A_5$ is an irreducible fifteen dimensional representation. Thus the embedding which maps the six into the six, also maps the fifteen into the fifteen.

It is clear that the approach above will not help us find the S-subalgebras of the exceptional Lie algebras. Fortunately, this problem has been solved, again by Dynkin. In order to display his results we must introduce some additional notation. In particular, we need a means of specifying a particular embedding of a subalgebra $G'$ in an algebra, $G$. This is done with the **index of the embedding**. In Chapter XI, we introduced the concept of the index of a representation, which is simply the ratio of the bilinear form obtained from the trace of the product of two representation matrices to the bilinear form which is the Killing form normalized in a particular way. Here we define the index of an embedding to be the ratio of the bilinear form on $G'$ obtained by lifting the value of the Killing form on $G$ to the Killing form on $G'$ itself:

$$j_f(x', y')'_2 = (f(x'), f(y'))_2 \tag{XV.16}$$

where $j_f$ is the index of the embedding and $f : G' \to G$ is the embedding. As we have seen, on a simple Lie algebra all invariant bilinear forms are proportional, so this definition makes sense. Now suppose $\phi$ is a representation of G, that is, a mapping of elements of $G$ onto a space of linear transformations. Then $\phi \circ f$ is a representation of $G'$. Moreover, for $x', y' \in G'$:

$$Tr\ \phi(f(x'))\phi(f(y')) = l_{\phi \circ f}(x', y')'_2$$

$$= l_\phi(f(x')), f(y'))_2$$

$$= l_\phi j_f(x', y')'_2 \tag{XV.17}$$

so we see that the index of the embedding is determined by the ratio of the indices of the representations $\phi$ and $\phi \circ f$:

$$j_f = \frac{l_{\phi \circ f}}{l_\phi} . \tag{XV.18}$$

Consider $G_2$, which we know has a 7 dimensional representation. Thus we might hope to embed $A_1$, which has a 7 dimensional representation, in $G_2$. This is in fact possible. Now we compute the index of the seven dimensional representation of $A_1$ according to the methods of Chapter XI: $l = 6 \times 7 \times 8/6 = 56$. For the seven dimensional representation of $G_2$ we have $l = 7\langle \Lambda, \Lambda + 2\delta \rangle / 14 = 2$. See the Problems following Chapter XI. Thus the index of the embedding is 28. Dynkin indicates this subalgebra by $A_1^{28}$. If there is more than one subalgebra with the same index, we can use primes to indicate this.

Having established this notation, we list the results of Dynkin[3] for the maximal S-subalgebras of the exceptional Lie algebras:

Maximal S-subalgebras of Exceptional Lie Algebras

$G_2$   $A_1^{28}$

$F_4$   $A_1^{156}, G_2^1 + A_1^8$

$E_6$   $A_2^9, G_2^3, C_4^1, G_2^1 + A_2^{2\prime\prime}, F_4^1$

$E_7$   $A_1^{399}, A_1^{231}, A_2^{21}, G_2^1 + C_3^{1\prime\prime}, F_4^1 + A_1^{3\prime\prime}, G_2^2 + A_1^7, A_1^{24} + A_1^{15}$

$E_8$   $A_1^{1240}, A_1^{760}, A_1^{520}, G_2^1 + F_4^1, A_2^{6\prime} + A_1^{16}, B_2^{12}$

We summarize here the results on the maximal semi-simple subalgebras of the simple Lie algebras:

1. Regular subalgebras are found using the algorithm of extended Dynkin diagrams.

   a. Dropping a dot from an extended diagram yields a regular subalgebra which is semi-simple and maximal unless it is one of the exceptions mentioned on pp. 134 - 135.

   b. Dropping a dot from a basic diagram yields a subalgebra which may be maximal among the semi-simple subalgebras.

2. Non-Regular (S-subalgebras)

   a. Of classical algebras:

   i. Non-simple: $SU(s) \times SU(t) \subset SU(st)$, $SO(s) \times SO(t) \subset SO(st)$, $Sp(s) \times Sp(t) \subset SO(st)$, $Sp(s) \times SO(t) \subset Sp(st)$ , and $O(s) \times O(t) \subset O(s+t)$ for $s$ and $t$ odd.

   ii. Simple: If $G$ has an n dimensional representation it is maximal in $SU(n)$, $SO(n)$, or $Sp(n)$, unless it is one of the few exceptions listed by Dynkin. If the representation has a symmetric bilinear form the subalgebra is maximal in $SO(n)$. If it has an anti-symmetric bilinear form, it is maximal in $Sp(n)$. If it has no bilinear form, it is maximal in $SU(n)$.

   b. Of exceptional Lie algebras: the maximal S-subalgebras are listed above.

## Footnotes

1. DYNKIN II, p. 148.

2. GOLUBITSKY and ROTHSCHILD.

3. DYNKIN III.

4. DYNKIN III, p. 365.

5. DYNKIN II, p. 231.

## References

This material is comes from DYNKIN II, III.

Very useful tables are provided in SLANSKY.

Incredibly extensive tables are given in MC KAY and PATERA.

## Exercises

1. Find the maximal semi-simple subalgebras of $A_4$.

2. Find the maximal semi-simple subalgebras of $D_5$. Note that it is necessary to consider some subalgebras which are only maximal among the semi-simple subalgebras. ans. $A_3 + A_1 + A_1, A_4, B_4, D_4, A_1 + B_2, B_2 + B_2$.

3. Find the maximal semi-simple subalgebras of $F_4$.

4. Find the maximal semi-simple subalgebras of $B_2$.

# XVI. Branching Rules

Having determined, with much help from E. B. Dynkin, the maximal semi-simple subalgebras of the simple Lie algebras, we want to pursue this further to learn how an irreducible representation of an algebra becomes a representation of the subalgebra. To do this we shall have to be more precise about the embedding of the subalgebra in the algebra. Indeed, as we have already seen, the three dimensional representation of $SU(3)$ may become either a reducible representation or an irreducible representation of $SU(2)$ depending on the embedding.

We start with a subalgebra $G'$ embedded in the algebra $G$ by a mapping $f : G' \to G$, where $f$ is a homomorphism, that is, it preserves the commutation relations:

$$f([x', y']) = [f(x'), f(y')], \qquad x', y' \in G' . \qquad \text{(XVI.1)}$$

Moreover, we can arrange it so that the Cartan subalgebra $H' \subset G'$ is mapped by $f$ into the Cartan subalgebra $H \subset G$. Note that if $\phi$ is a representation of $G$, then

$\phi \circ f$ is a representation of $G'$.

If we are to make progress, we must deal not only with the algebras but with the root spaces $H_0^*$ $'$and $H_0^*$ as well. Given the mapping $f$, we define $f^* : H_0^* \to H_0^*$ $'$ by

$$f^* \circ \rho = \rho \circ f \qquad \qquad \text{(XVI.2)}$$

where $\rho \in H_0^*$. That is to say, if $h' \in H'$

$$(f^* \circ \rho)(h') = \rho(f(h')) . \qquad \qquad \text{(XVI.3)}$$

Instead of thinking of $G'$ as external to $G$, it is easier to imagine it already within $G$. Then $f : H' \to H$ simply maps $H'$ onto itself as the identity. We recall that there is a one-to-one correspondence between the elements of the root space, $\rho \in H_0^*$ and the elements $h_\rho$ of the Cartan subalgebra. This one-to-one mapping connects to $H'$ a space which we regard as $H_0^*$ $'$. Now let us decompose $H_0^*$ as the sum of $H_0^*$ $'$ and a space $H^*$ orthogonal to it. That is, if $\tau \in H_0^*$ $'$ and $\kappa \in H^*$, then $\langle \tau, \kappa \rangle = 0$. Then the elements of $H^*$ are functionals which when applied to $H'$ give zero. This is so because if $\kappa \in H^*$ corresponds to $h_\kappa \in H$ and $\tau \in H_0^*$ $'$ corresponds to $h_\tau \in H'$, then $\kappa(h_\tau) = (h_\kappa, h_\tau) = \langle \kappa, \tau \rangle = 0$. Now the action of $f^*$ is simply to project elements of $H_0^*$ onto $H_0^*$ $'$. This follows because if $\rho = \rho_1 + \rho_2, \rho_1 \in H_0^*$ $', \rho_2 \in H^*$, then for $h' \in H'$, $f^* \circ \rho(h') = \rho(f(h')) = (\rho_1 + \rho_2)(h') = \rho_1(h')$. Thus $f^* \circ \rho = \rho_1$.

Let $M$ be a weight of a representation $\phi$ of $G$:

$$\phi(h)\xi_M = M(h)\xi_M . \qquad \qquad \text{(XVI.4)}$$

Then if $h \in H' \subset H$,

$$\phi(f(h))\xi_M = M(f(h))\xi_M$$

$$= f^* \circ M(h)\xi_M . \qquad \qquad \text{(XVI.5)}$$

It follows that if $M$ is a weight of $\phi$, then $f^* \circ M$ is a weight of the representation $\phi \circ f$ of $G'$. More graphically, the weights of the representation of the subalgebra are obtained by projecting the weights of the algebra from $H_0^*$ onto $H_0^*{}'$.

A simple but important consequence of this conclusion is that if the rank of $G'$ is the same as the rank of $G$, then the subalgebra is regular. This is so because if the ranks are the same, then $H_0^*{}'$ coincides with $H_0^*$ so the projection is simply the identity. Thus if we start with the adjoint representation of $G$, it will be mapped (by the identity) into a reducible representation of $G'$ which contains the adjoint of $G'$. But the weights of the adjoint of $G'$ must then have been among the weights of the adjoint of $G$.

Let us pause to consider an example. Let us take $G = G_2$ and $G' = A_2$. We know this is a regular subalgebra. Indeed, examining the root diagram, we see that the six long roots form the familiar hexagon of the adjoint representation of $SU(3) = A_2$. The projection $f^*$ here is just the identity map since the algebra and the subalgebra have the same rank. The fourteen dimensional adjoint representation becomes the sum of the eight dimensional adjoint representation of $SU(3)$ and two conjugate three dimensional representations.

We state without proof two of Dynkin's theorems.[1] If $G'$ is a regular subalgebra of $G$ and $\phi$ is representation of $G$, then $\phi \circ f$ is reducible. An approximate converse is also true. If $G$ is $A_n, B_n$ or $C_n$, and $G'$ has a reducible representation which makes it a subalgebra of $G$ by being respectively n-dimensional with no bilinear invariant or 2n+1 dimensional with a symmetric bilinear invariant, or 2n dimensional with an anti-symmetric bilinear invariant, then $G'$ is regular. In other words, if $A_n, B_n,$ or $C_n$ has an S-subalgebra, that S-subalgebra must have an irreducible representation of dimension, n, 2n+1, or 2n respectively.

What happened to $D_n$ in this theorem? As we saw in the last chapter, $O(s) \times O(t)$ is an S-subalgebra of $O(s + t)$ if both $s$ and $t$ are odd.

We now proceed to the determination of $f^*$, the mapping which connects $H_0^*$ to $H_0^*{}'$. Once we know $f^*$ we can find the weights of $\phi \circ f$ for any representation $\phi$ of a given algebra. From these weights we can infer the irreducible representations into which $\phi \circ f$ decomposes. In fact, extensive tables of these branching rules have been compiled by computer (see McKay and Patera). Here we seek to develop some intuitive understanding of the procedure.

To be explicit, we shall take an example: $B_3$, that is $O(7)$. By the methods of the previous chapter we can easily find the maximal regular semi-simple subalgebras: $A_3 = D_3$ and $A_1 + A_1 + A_1$. In seeking the S-subalgebras, we note that the technique $O(s) \times O(t) \subset O(st)$ is of no avail for $st = 7$. On the other hand, $A_1$ has a seven dimensional representation which has a symmetric bilinear form. Thus we might anticipate that $A_1$ is a maximal S-subalgebra of $B_3$. In fact, as we shall see, $A_1 \subset G_2$ and $G_2$ is maximal in $B_3$.

There is a simple and effective way to find the mappings $f^*$ for the regular subalgebras. Remember the procedure for constructing the extended Dynkin diagrams. We added the vector $-\gamma$ to the diagram for the simple algebra, where $\gamma$ was the highest root. We then had a diagram for the set which was the union of $-\gamma$ and the simple roots. From this set, we struck one root. The remainder furnished a basis of simple roots for the subalgebra. The Dynkin coefficients for the weights relative to the new basis of simple roots are just the Dynkin coefficients with respect to the surviving old simple roots, together with the Dynkin coefficient with respect to $-\gamma$. Thus we simply calculate the Dynkin coefficient of each weight with respect to $-\gamma$ and use it in place of one of the old coefficients. Calculating the coefficient with respect to $-\gamma$ is trivial since we can express $-\gamma$ as a linear combination of the simple roots.

Let us use this technique to analyze the regular subalgebras of $B_3$. The Dynkin coefficients of $\gamma$ are $(0, 1, 0)$. Referring to the Cartan matrix we see that $-\gamma = \alpha_1 + 2\alpha_2 + 2\alpha_3$. Since $\langle \gamma, \gamma \rangle = \langle \alpha_1, \alpha_1 \rangle = \langle \alpha_2, \alpha_2 \rangle = 2\langle \alpha_3, \alpha_3 \rangle$, the coefficient of a weight with respect to $-\gamma$ is $-a_1 - 2a_2 - a_3$ if the Dynkin coefficients with respect to $\alpha_1, \alpha_2$, and $\alpha_3$ are $a_1, a_2$, and $a_3$ respectively.

In this way we construct the **extended weight scheme** for the seven dimensional representation of $B_3$:

$$
\begin{array}{cccc}
1 & 0 & 0 & -1 \\
\end{array}
$$

$$
\begin{array}{cccc}
-1 & 1 & 0 & -1 \\
\end{array}
$$

$$
\begin{array}{cccc}
0 & -1 & 2 & 0 \\
\end{array}
$$

$$
\begin{array}{cccc}
0 & 0 & 0 & 0 \\
\end{array}
$$

$$
\begin{array}{cccc}
0 & 1 & -2 & 0 \\
\end{array}
$$

$$
\begin{array}{cccc}
1 & -1 & 0 & 1 \\
\end{array}
$$

$$
\begin{array}{cccc}
-1 & 0 & 0 & 1 \\
\end{array}
$$

Now the $A_3$ regular subalgebra is obtained by using the fourth column, the one for $-\gamma$, rather than the third one. Deleting the third column we have

$$
\begin{array}{ccc}
1 & 0 & -1 \\
\end{array}
$$

$$
\begin{array}{ccc}
-1 & 1 & -1 \\
\end{array}
$$

$$
\begin{array}{ccc}
0 & -1 & 0 \\
\end{array}
$$

$$
\begin{array}{ccc}
0 & 0 & 0 \\
\end{array}
$$

$$
\begin{array}{ccc}
0 & 1 & 0 \\
\end{array}
$$

$$
\begin{array}{ccc}
1 & -1 & 1 \\
\end{array}
$$

$$
\begin{array}{ccc}
-1 & 0 & 1 \\
\end{array}
$$

This is a representation of $A_3$. The candidates for highest weights of irreducible components are $(0,1,0)$ and $(0,0,0)$ since these are the only ones with purely nonnegative Dynkin coefficients. Indeed, these give a six dimensional representation and a one-dimensional representation. Moreover, we can deduce the projection operator for this subalgebra directly by comparing three weights in the original basis to their values in the basis for the subalgebra $A_3$. In this way we find

$$f^*_{A_3}(1,0,0) = (1,0,-1)$$

$$f^*_{A_3}(0,1,0) = (0,1,-2)$$

$$f^*_{A_3}(0,0,1) = (0,0,-1) \; . \qquad \text{(XVI.6)}$$

Knowing this mapping gives us an alternative method of finding the weights of some irreducible representation of $B_3$ with respect to the subalgebra $A_3$. Thus for example, we can map the weights of the adjoint representation (it suffices just to consider the positive roots ) of $B_3$ into weights of $A_3$:

$$
\begin{array}{ccc}
\boxed{\begin{matrix} 0 & 1 & 0 \end{matrix}} & \rightarrow & \boxed{\begin{matrix} 0 & 1 & -2 \end{matrix}} \\[6pt]
\boxed{\begin{matrix} 1 & -1 & 2 \end{matrix}} & \rightarrow & \boxed{\begin{matrix} 1 & -1 & -1 \end{matrix}} \\[6pt]
\boxed{\begin{matrix} -1 & 0 & 2 \end{matrix}} & \rightarrow & \boxed{\begin{matrix} -1 & 0 & -1 \end{matrix}} \\[6pt]
\boxed{\begin{matrix} 1 & 0 & 0 \end{matrix}} & \rightarrow & \boxed{\begin{matrix} 1 & 0 & -1 \end{matrix}} \\[6pt]
\boxed{\begin{matrix} -1 & 1 & 0 \end{matrix}} & \rightarrow & \boxed{\begin{matrix} -1 & 1 & -1 \end{matrix}} \\[6pt]
\boxed{\begin{matrix} 1 & 1 & -2 \end{matrix}} & \rightarrow & \boxed{\begin{matrix} 1 & 1 & -1 \end{matrix}} \\[6pt]
\boxed{\begin{matrix} 0 & -1 & 2 \end{matrix}} & \rightarrow & \boxed{\begin{matrix} 0 & -1 & 0 \end{matrix}} \\[6pt]
\boxed{\begin{matrix} -1 & 2 & -2 \end{matrix}} & \rightarrow & \boxed{\begin{matrix} -1 & 2 & -1 \end{matrix}} \\[6pt]
\boxed{\begin{matrix} 2 & -1 & 0 \end{matrix}} & \rightarrow & \boxed{\begin{matrix} 2 & -1 & 0 \end{matrix}}
\end{array}
$$

From these weights and their negatives, it is clear that the candidates for highest weights are $(1,0,1)$ and $(0,1,0)$ which indeed correspond to representations of dimension fifteen and six respectively.

If we consider the subalgebra $A_1 + A_1 + A_1$, deleting the Dynkin coefficients with respect to the second root, we find the seven dimensional representation of $B_3$ is mapped into

| 1 | 0 | −1 |
|---|---|---|

| −1 | 0 | −1 |
|---|---|---|

| 0 | 2 | 0 |
|---|---|---|

| 0 | 0 | 0 |
|---|---|---|

| 0 | −2 | 0 |
|---|---|---|

| 1 | 0 | 1 |
|---|---|---|

| −1 | 0 | 1 |
|---|---|---|

This is the reducible representation whose Dynkin expression is $[(0)+(2)+(0)] + [(1)+(0)+(1)]$. In the notation which indicates dimensionality, it is $(1,3,1)+(2,1,2)$.

It is clear that the regular subalgebras can be dealt with in a very simple fashion. The S-subalgebras require more effort. It is always possible to order the Cartan subalgebras of the initial algebra and of its subalgebra so that if $x > y$, $x, y \in H'$ then $f^*(x) > f^*(y)$. Now we exploit this by writing the weights of the seven dimensional representations of $B_3$ and $G_2$ beside each other:

$$
\begin{array}{|ccc|}\hline 1 & 0 & 0 \\ \hline\end{array} \qquad \begin{array}{|cc|}\hline 0 & 1 \\ \hline\end{array}
$$

$$
\begin{array}{|ccc|}\hline -1 & 1 & 0 \\ \hline\end{array} \qquad \begin{array}{|cc|}\hline 1 & -1 \\ \hline\end{array}
$$

$$
\begin{array}{|ccc|}\hline 0 & -1 & 2 \\ \hline\end{array} \qquad \begin{array}{|cc|}\hline -1 & 2 \\ \hline\end{array}
$$

$$
\begin{array}{|ccc|}\hline 0 & 0 & 0 \\ \hline\end{array} \qquad \begin{array}{|cc|}\hline 0 & 0 \\ \hline\end{array}
$$

$$
\begin{array}{|ccc|}\hline 0 & 1 & -2 \\ \hline\end{array} \qquad \begin{array}{|cc|}\hline 1 & -2 \\ \hline\end{array}
$$

$$
\begin{array}{|ccc|}\hline 1 & -1 & 0 \\ \hline\end{array} \qquad \begin{array}{|cc|}\hline -1 & 1 \\ \hline\end{array}
$$

$$
\begin{array}{|ccc|}\hline -1 & 0 & 0 \\ \hline\end{array} \qquad \begin{array}{|cc|}\hline 0 & -1 \\ \hline\end{array}
$$

Thus it is clear that we must have

$$f^*_{G_2}(1,0,0) = (0,1)$$

$$f^*_{G_2}(0,1,0) = (1,0)$$

$$f^*_{G_2}(0,0,1) = (0,1) \ . \tag{XVI.7}$$

Equipped with this mapping, we can project the weights of any representation of $B_3$ onto the space of weights of $G_2$ and thus identify the branching rules.

For example, we again consider the positive roots of $B_3$ to find the branching rule for the adjoint representation:

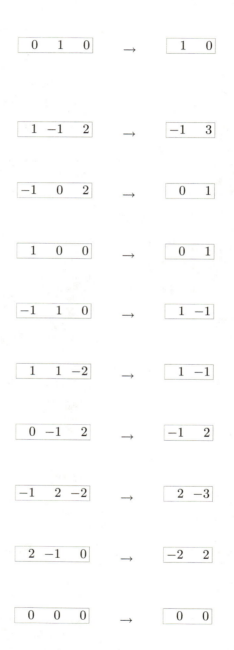

The weights with non-negative Dynkin coefficients are (1,0), (0,1), and (0,1). Now the fourteen dimensional representation has highest weight (1,0) and includes the weight (0,1). Thus we see that the 21 dimensional representation of $B_3$ becomes the sum of a 14 dimensional representation and a 7 dimensional representation.

## Footnote

1. DYNKIN III, pp. 158–159.

## References

Again, the entire chapter is due to DYNKIN III. The works of SLANSKY and of MC KAY AND PATERA provide exhaustive tables.

## Exercises

1. Find the branching rules for the ten-dimensional representation of $SU(5)$ for the maximal semi-simple subalgebras.

2. Find the branching rules for the 26 and 52 dimensional representations of $F_4$ into its maximal semi-simple subalgebras.

# Bibliography

BOERNER, H., *Representations of Groups*, North-Holland, Amsterdam, 1970.

CARRUTHERS, P., *Introduction to Unitary Symmetry*, Interscience, New York, 1966.

DYNKIN, E. B., The structure of semi-simple Lie algebras, *Uspehi Mat. Nauk* **2** (1947) pp. 59–127. *Am. Math. Soc. Trans. No. 1* (1950) pp. 1–143. (DYNKIN I)

DYNKIN, E. B., Semi-simple subalgebras of semi-simple Lie algebras, *Mat. Sbornik* **30**(1952) pp. 349–462. *Am. Math. Soc. Trans. Ser. 2*, **6** (1957) pp. 111–244. (DYNKIN II)

DYNKIN, E. B. Maximal subgroups of the classical groups. *Trudy Moskov. Mat. Obshchestva* **1**(1952), pp. 39–166. *Am. Math. Soc. Trans. Ser. 2*, **6** (1957) pp. 245–378. (DYNKIN III)

GASIOROWICZ, S. *Elementary Particle Physics*, John Wiley & Sons., New York, 1966.

GEORGI, H., *Lie Algebras in Particle Physics*, Benjamin/Cummings, Reading, Mass. 1982.

GILMORE, R., *Lie Groups, Lie Algebras, and Some of their Applications* , John Wiley & Sons, New York, 1974.

GOLUBITSKY, M. and ROTHSCHILD, B., Primitive Subalgebras of Exceptional Lie Algebras, *Bull. Am. Math. Soc.* **77** (1971) pp. 983–986.

GOTTFRIED, K.,*Quantum Mechanics, v. I*, W. A. Benjamin, New York, 1966.

HAMERMESH, M.,*Group Theory and its Application to Physical Problems*, Addison-Wesley, Reading, 1962.

JACOBSON, N., *Lie Algebras*, Interscience, 1962; Dover, New York, 1979.

MC KAY, W.B., and PATERA, J., *Tables of Dimensions, Indices, and Branching Rules for Representations of Simple Lie Algebras*, Marcel Dekker, New York, 1981.

MILLER, W., *Symmetry Groups and their Applications*, Academic Press, New York, 1972.

SCHENSTED, I., *A Course on the Applications of Group Theory to Quantum Mechanics*, NEO Press, Peaks Is., Me. 1976.

SLANSKY, R., Lie Algebras for Model Building, *Physics Reports* **79** (1981) 1.

# Index

A CATALOG OF SELECTED
DOVER BOOKS
IN SCIENCE AND MATHEMATICS

# Mathematics

FUNCTIONAL ANALYSIS (Second Corrected Edition), George Bachman and Lawrence Narici. Excellent treatment of subject geared toward students with background in linear algebra, advanced calculus, physics and engineering. Text covers introduction to inner-product spaces, normed, metric spaces, and topological spaces; complete orthonormal sets, the Hahn-Banach Theorem and its consequences, and many other related subjects. 1966 ed. 544pp. 6⅛ x 9¼. 0-486-40251-7

ASYMPTOTIC EXPANSIONS OF INTEGRALS, Norman Bleistein & Richard A. Handelsman. Best introduction to important field with applications in a variety of scientific disciplines. New preface. Problems. Diagrams. Tables. Bibliography. Index. 448pp. 5⅜ x 8½. 0-486-65082-0

VECTOR AND TENSOR ANALYSIS WITH APPLICATIONS, A. I. Borisenko and I. E. Tarapov. Concise introduction. Worked-out problems, solutions, exercises. 257pp. 5⅜ x 8¼. 0-486-63833-2

AN INTRODUCTION TO ORDINARY DIFFERENTIAL EQUATIONS, Earl A. Coddington. A thorough and systematic first course in elementary differential equations for undergraduates in mathematics and science, with many exercises and problems (with answers). Index. 304pp. 5⅜ x 8½. 0-486-65942-9

FOURIER SERIES AND ORTHOGONAL FUNCTIONS, Harry F. Davis. An incisive text combining theory and practical example to introduce Fourier series, orthogonal functions and applications of the Fourier method to boundary-value problems. 570 exercises. Answers and notes. 416pp. 5⅜ x 8½. 0-486-65973-9

COMPUTABILITY AND UNSOLVABILITY, Martin Davis. Classic graduate-level introduction to theory of computability, usually referred to as theory of recurrent functions. New preface and appendix. 288pp. 5⅜ x 8½. 0-486-61471-9

ASYMPTOTIC METHODS IN ANALYSIS, N. G. de Bruijn. An inexpensive, comprehensive guide to asymptotic methods—the pioneering work that teaches by explaining worked examples in detail. Index. 224pp. 5⅜ x 8½ 0-486-64221-6

APPLIED COMPLEX VARIABLES, John W. Dettman. Step-by-step coverage of fundamentals of analytic function theory—plus lucid exposition of five important applications: Potential Theory; Ordinary Differential Equations; Fourier Transforms; Laplace Transforms; Asymptotic Expansions. 66 figures. Exercises at chapter ends. 512pp. 5⅜ x 8½. 0-486-64670-X

INTRODUCTION TO LINEAR ALGEBRA AND DIFFERENTIAL EQUATIONS, John W. Dettman. Excellent text covers complex numbers, determinants, orthonormal bases, Laplace transforms, much more. Exercises with solutions. Undergraduate level. 416pp. 5⅜ x 8½. 0-486-65191-6

RIEMANN'S ZETA FUNCTION, H. M. Edwards. Superb, high-level study of landmark 1859 publication entitled "On the Number of Primes Less Than a Given Magnitude" traces developments in mathematical theory that it inspired. xiv+315pp. 5⅜ x 8½. 0-486-41740-9

CALCULUS OF VARIATIONS WITH APPLICATIONS, George M. Ewing. Applications-oriented introduction to variational theory develops insight and promotes understanding of specialized books, research papers. Suitable for advanced undergraduate/graduate students as primary, supplementary text. 352pp. 5⅜ x 8½.
0-486-64856-7

COMPLEX VARIABLES, Francis J. Flanigan. Unusual approach, delaying complex algebra till harmonic functions have been analyzed from real variable viewpoint. Includes problems with answers. 364pp. 5⅜ x 8½. 0-486-61388-7

AN INTRODUCTION TO THE CALCULUS OF VARIATIONS, Charles Fox. Graduate-level text covers variations of an integral, isoperimetrical problems, least action, special relativity, approximations, more. References. 279pp. 5⅜ x 8½.
0-486-65499-0

COUNTEREXAMPLES IN ANALYSIS, Bernard R. Gelbaum and John M. H. Olmsted. These counterexamples deal mostly with the part of analysis known as "real variables." The first half covers the real number system, and the second half encompasses higher dimensions. 1962 edition. xxiv+198pp. 5⅜ x 8½. 0-486-42875-3

CATASTROPHE THEORY FOR SCIENTISTS AND ENGINEERS, Robert Gilmore. Advanced-level treatment describes mathematics of theory grounded in the work of Poincaré, R. Thom, other mathematicians. Also important applications to problems in mathematics, physics, chemistry and engineering. 1981 edition. References. 28 tables. 397 black-and-white illustrations. xvii + 666pp. 6⅛ x 9¼.
0-486-67539-4

INTRODUCTION TO DIFFERENCE EQUATIONS, Samuel Goldberg. Exceptionally clear exposition of important discipline with applications to sociology, psychology, economics. Many illustrative examples; over 250 problems. 260pp. 5⅜ x 8½.
0-486-65084-7

NUMERICAL METHODS FOR SCIENTISTS AND ENGINEERS, Richard Hamming. Classic text stresses frequency approach in coverage of algorithms, polynomial approximation, Fourier approximation, exponential approximation, other topics. Revised and enlarged 2nd edition. 721pp. 5⅜ x 8½. 0-486-65241-6

INTRODUCTION TO NUMERICAL ANALYSIS (2nd Edition), F. B. Hildebrand. Classic, fundamental treatment covers computation, approximation, interpolation, numerical differentiation and integration, other topics. 150 new problems. 669pp. 5⅜ x 8½. 0-486-65363-3

THREE PEARLS OF NUMBER THEORY, A. Y. Khinchin. Three compelling puzzles require proof of a basic law governing the world of numbers. Challenges concern van der Waerden's theorem, the Landau-Schnirelmann hypothesis and Mann's theorem, and a solution to Waring's problem. Solutions included. 64pp. 5⅜ x 8½.
0-486-40026-3

THE PHILOSOPHY OF MATHEMATICS: AN INTRODUCTORY ESSAY, Stephan Körner. Surveys the views of Plato, Aristotle, Leibniz & Kant concerning propositions and theories of applied and pure mathematics. Introduction. Two appendices. Index. 198pp. 5⅜ x 8½. 0-486-25048-2

INTRODUCTORY REAL ANALYSIS, A.N. Kolmogorov, S. V. Fomin. Translated by Richard A. Silverman. Self-contained, evenly paced introduction to real and functional analysis. Some 350 problems. 403pp. 5⅜ x 8½.                    0-486-61226-0

APPLIED ANALYSIS, Cornelius Lanczos. Classic work on analysis and design of finite processes for approximating solution of analytical problems. Algebraic equations, matrices, harmonic analysis, quadrature methods, much more. 559pp. 5⅜ x 8½.
0-486-65656-X

AN INTRODUCTION TO ALGEBRAIC STRUCTURES, Joseph Landin. Superb self-contained text covers "abstract algebra": sets and numbers, theory of groups, theory of rings, much more. Numerous well-chosen examples, exercises. 247pp. 5⅜ x 8½.
0-486-65940-2

QUALITATIVE THEORY OF DIFFERENTIAL EQUATIONS, V. V. Nemytskii and V.V. Stepanov. Classic graduate-level text by two prominent Soviet mathematicians covers classical differential equations as well as topological dynamics and ergodic theory. Bibliographies. 523pp. 5⅜ x 8½.                    0-486-65954-2

THEORY OF MATRICES, Sam Perlis. Outstanding text covering rank, nonsingularity and inverses in connection with the development of canonical matrices under the relation of equivalence, and without the intervention of determinants. Includes exercises. 237pp. 5⅜ x 8½.                    0-486-66810-X

INTRODUCTION TO ANALYSIS, Maxwell Rosenlicht. Unusually clear, accessible coverage of set theory, real number system, metric spaces, continuous functions, Riemann integration, multiple integrals, more. Wide range of problems. Undergraduate level. Bibliography. 254pp. 5⅜ x 8½.                    0-486-65038-3

MODERN NONLINEAR EQUATIONS, Thomas L. Saaty. Emphasizes practical solution of problems; covers seven types of equations. ". . . a welcome contribution to the existing literature...."—*Math Reviews*. 490pp. 5⅜ x 8½.                    0-486-64232-1

MATRICES AND LINEAR ALGEBRA, Hans Schneider and George Phillip Barker. Basic textbook covers theory of matrices and its applications to systems of linear equations and related topics such as determinants, eigenvalues and differential equations. Numerous exercises. 432pp. 5⅜ x 8½.                    0-486-66014-1

LINEAR ALGEBRA, Georgi E. Shilov. Determinants, linear spaces, matrix algebras, similar topics. For advanced undergraduates, graduates. Silverman translation. 387pp. 5⅜ x 8½.                    0-486-63518-X

ELEMENTS OF REAL ANALYSIS, David A. Sprecher. Classic text covers fundamental concepts, real number system, point sets, functions of a real variable, Fourier series, much more. Over 500 exercises. 352pp. 5⅜ x 8½.                    0-486-65385-4

SET THEORY AND LOGIC, Robert R. Stoll. Lucid introduction to unified theory of mathematical concepts. Set theory and logic seen as tools for conceptual understanding of real number system. 496pp. 5⅜ x 8¼.                    0-486-63829-4

# Math–Decision Theory, Statistics, Probability

ELEMENTARY DECISION THEORY, Herman Chernoff and Lincoln E. Moses. Clear introduction to statistics and statistical theory covers data processing, probability and random variables, testing hypotheses, much more. Exercises. 364pp. 5⅜ x 8½. 0-486-65218-1

STATISTICS MANUAL, Edwin L. Crow et al. Comprehensive, practical collection of classical and modern methods prepared by U.S. Naval Ordnance Test Station. Stress on use. Basics of statistics assumed. 288pp. 5⅜ x 8½. 0-486-60599-X

SOME THEORY OF SAMPLING, William Edwards Deming. Analysis of the problems, theory and design of sampling techniques for social scientists, industrial managers and others who find statistics important at work. 61 tables. 90 figures. xvii +602pp. 5⅜ x 8½. 0-486-64684-X

LINEAR PROGRAMMING AND ECONOMIC ANALYSIS, Robert Dorfman, Paul A. Samuelson and Robert M. Solow. First comprehensive treatment of linear programming in standard economic analysis. Game theory, modern welfare economics, Leontief input-output, more. 525pp. 5⅜ x 8½. 0-486-65491-5

PROBABILITY: AN INTRODUCTION, Samuel Goldberg. Excellent basic text covers set theory, probability theory for finite sample spaces, binomial theorem, much more. 360 problems. Bibliographies. 322pp. 5⅜ x 8½. 0-486-65252-1

GAMES AND DECISIONS: INTRODUCTION AND CRITICAL SURVEY, R. Duncan Luce and Howard Raiffa. Superb nontechnical introduction to game theory, primarily applied to social sciences. Utility theory, zero-sum games, n-person games, decision-making, much more. Bibliography. 509pp. 5⅜ x 8½. 0-486-65943-7

INTRODUCTION TO THE THEORY OF GAMES, J. C. C. McKinsey. This comprehensive overview of the mathematical theory of games illustrates applications to situations involving conflicts of interest, including economic, social, political, and military contexts. Appropriate for advanced undergraduate and graduate courses; advanced calculus a prerequisite. 1952 ed. x+372pp. 5⅜ x 8½. 0-486-42811-7

FIFTY CHALLENGING PROBLEMS IN PROBABILITY WITH SOLUTIONS, Frederick Mosteller. Remarkable puzzlers, graded in difficulty, illustrate elementary and advanced aspects of probability. Detailed solutions. 88pp. 5⅜ x 8½. 65355-2

PROBABILITY THEORY: A CONCISE COURSE, Y. A. Rozanov. Highly readable, self-contained introduction covers combination of events, dependent events, Bernoulli trials, etc. 148pp. 5⅜ x 8¼. 0-486-63544-9

STATISTICAL METHOD FROM THE VIEWPOINT OF QUALITY CONTROL, Walter A. Shewhart. Important text explains regulation of variables, uses of statistical control to achieve quality control in industry, agriculture, other areas. 192pp. 5⅜ x 8½. 0-486-65232-7

TENSOR CALCULUS, J.L. Synge and A. Schild. Widely used introductory text covers spaces and tensors, basic operations in Riemannian space, non-Riemannian spaces, etc. 324pp. 5⅜ x 8¼. 0-486-63612-7

ORDINARY DIFFERENTIAL EQUATIONS, Morris Tenenbaum and Harry Pollard. Exhaustive survey of ordinary differential equations for undergraduates in mathematics, engineering, science. Thorough analysis of theorems. Diagrams. Bibliography. Index. 818pp. 5⅜ x 8½. 0-486-64940-7

INTEGRAL EQUATIONS, F. G. Tricomi. Authoritative, well-written treatment of extremely useful mathematical tool with wide applications. Volterra Equations, Fredholm Equations, much more. Advanced undergraduate to graduate level. Exercises. Bibliography. 238pp. 5⅜ x 8½. 0-486-64828-1

FOURIER SERIES, Georgi P. Tolstov. Translated by Richard A. Silverman. A valuable addition to the literature on the subject, moving clearly from subject to subject and theorem to theorem. 107 problems, answers. 336pp. 5⅜ x 8½. 0-486-63317-9

INTRODUCTION TO MATHEMATICAL THINKING, Friedrich Waismann. Examinations of arithmetic, geometry, and theory of integers; rational and natural numbers; complete induction; limit and point of accumulation; remarkable curves; complex and hypercomplex numbers, more. 1959 ed. 27 figures. xii+260pp. 5⅜ x 8½. 0-486-63317-9

POPULAR LECTURES ON MATHEMATICAL LOGIC, Hao Wang. Noted logician's lucid treatment of historical developments, set theory, model theory, recursion theory and constructivism, proof theory, more. 3 appendixes. Bibliography. 1981 edition. ix + 283pp. 5⅜ x 8½. 0-486-67632-3

CALCULUS OF VARIATIONS, Robert Weinstock. Basic introduction covering isoperimetric problems, theory of elasticity, quantum mechanics, electrostatics, etc. Exercises throughout. 326pp. 5⅜ x 8½. 0-486-63069-2

THE CONTINUUM: A CRITICAL EXAMINATION OF THE FOUNDATION OF ANALYSIS, Hermann Weyl. Classic of 20th-century foundational research deals with the conceptual problem posed by the continuum. 156pp. 5⅜ x 8½. 0-486-67982-9

CHALLENGING MATHEMATICAL PROBLEMS WITH ELEMENTARY SOLUTIONS, A. M. Yaglom and I. M. Yaglom. Over 170 challenging problems on probability theory, combinatorial analysis, points and lines, topology, convex polygons, many other topics. Solutions. Total of 445pp. 5⅜ x 8½. Two-vol. set. Vol. I: 0-486-65536-9 Vol. II: 0-486-65537-7

Paperbound unless otherwise indicated. Available at your book dealer, online at **www.doverpublications.com**, or by writing to Dept. GI, Dover Publications, Inc., 31 East 2nd Street, Mineola, NY 11501. For current price information or for free catalogues (please indicate field of interest), write to Dover Publications or log on to **www.doverpublications.com** and see every Dover book in print. Dover publishes more than 500 books each year on science, elementary and advanced mathematics, biology, music, art, literary history, social sciences, and other areas.